01-HHE-2a

AF621382

MOLECULAR ALTERNATIVES TO INDICATOR AND PATHOGEN DETECTION: REAL-TIME PCR

by:

Weiping Chu
Sam Choi
Hojabr Dezfulian
Mandy Han
Jian-Wen He
Sunny Jiang

Department of Environmental Health, Science and Policy
University of California Irvine

2006

WERF

The Water Environment Research Foundation, a not-for-profit organization, funds and manages water quality research for its subscribers through a diverse public-private partnership between municipal utilities, corporations, academia, industry, and the federal government. WERF subscribers include municipal and regional water and wastewater utilities, industrial corporations, environmental engineering firms, and others that share a commitment to cost-effective water quality solutions. WERF is dedicated to advancing science and technology addressing water quality issues as they impact water resources, the atmosphere, the lands, and quality of life.

For more information, contact:
Water Environment Research Foundation
635 Slaters Lane, Suite 300
Alexandria, VA 22314-1177
Tel: (703) 684-2470
Fax: (703) 299-0742
www.werf.org
werf@werf.org

This report was co-published by the following organizations. For non-subscriber sales information, contact:

IWA Publishing
Alliance House, 12 Caxton Street
London SW1H 0QS, United Kingdom
Tel: +44 (0) 20 7654 5500
Fax: +44 (0) 20 7654 5555
www.iwapublishing.com
publications@iwap.co.uk

© Copyright 2006 by the Water Environment Research Foundation. All rights reserved. Permission to copy must be obtained from the Water Environment Research Foundation.
Library of Congress Catalog Card Number: 2005920223

IWAP ISBN: 978-1-84339-735-9

This report was prepared by the organization(s) named below as an account of work sponsored by the Water Environment Research Foundation (WERF). Neither WERF, members of WERF, the organization(s) named below, nor any person acting on their behalf: (a) makes any warranty, express or implied, with respect to the use of any information, apparatus, method, or process disclosed in this report or that such use may not infringe on privately owned rights; or (b) assumes any liabilities with respect to the use of, or for damages resulting from the use of, any information, apparatus, method, or process disclosed in this report.

University of California Irvine

This document was reviewed by a panel of independent experts selected by WERF. Mention of trade names or commercial products does not constitute WERF nor EPA endorsement or recommendations for use. Similarly, omission of products or trade names indicates nothing concerning WERF's nor EPA's positions regarding product effectiveness or applicability.

The research on which this report is based was funded, in part, by the United States Environmental Protection Agency through Cooperative Agreement No. CR-827345-01 with the Water Environment Research Foundation (WERF). Unless an EPA logo appears on the cover, this report is a publication of WERF, not EPA. Funds awarded under the Cooperative Agreement cited above were not used for editorial services, reproduction, printing, or distribution.

ACKNOWLEDGMENTS

The project team would like to acknowledge and thank the staff of Orange County Sanitation District for their support of this project.

Report Preparation

Principal Investigator:
Sunny Jiang, Ph.D.
University of California Irvine

Project Team:
Jian-Wen He, Ph.D.
Hojabr Dezfulian, Ph.D.
Yuli Tsai, Ph.D.
Weiping Chu
Mandy Han
Sam Choi
University of California Irvine

Charles McGee, M.S.
Orange County Sanitation District

Project Subcommittee

Suresh D. Pillai, Ph.D., Chair
Texas A&M University

Annette Prescott, Ph.D.
Thames Water Utilities Ltd.

Jorge Santo Domingo, Ph.D.
U.S. Environmental Protection Agency – NRMRL

Shawn Thompson, Ph.D.
Los Angeles County Sanitation Districts
San Jose Creek Water Quality Laboratory

Gary Toranzos, Ph.D.
University of Puerto Rico

Water Environmental Research Foundation Stuff

Director of Research: Daniel M. Woltering, Ph.D.
Program Manager: Lola Olabode

ABSTRACT AND BENEFITS

Abstract:

Real-time PCR methods were developed to detect enterococci and human adenoviruses in environmental samples collected from sewage and surface waters in southern California. The assays proved useful for detecting enterococci and human adenovirus in various environmental matrices including creek water, estuarine water, seawater and secondary sewage effluent. However, real-time PCR overestimated enterococci concentration in chlorinated secondary sewage and coastal water samples, suggesting non-infectious and viable but noncultivable bacteria are detectable by real-time PCR. Real-time PCR measurements of adenoviruses in urban rivers detected up to 10^4 adenovirus genome equivalence per liter of water sample. However, none of these PCR-positive samples, which were collected from the Los Angeles and San Gabriel Rivers, tested positive for infectious human adenoviruses using tissue culture plaque assay. Immunomagnetic capture of adenoviruses—using paramagnetic beads coated three monoclonal antibodies and one multicolonal antibody—was unsatisfactory due to non-specific binding of viruses to uncoated beads. Infectious human adenoviruses were detected in samples from Newport Bay, a small estuary surrounded by a highly urbanized area, using a novel HEK-293A cell line. This cell line appears to be more sensitive and supportive for the replication of adenovirus 40 than A549 cell line. Adenovirus hexon gene sequences retrieved from environmental isolates confirmed the presence of human adenovirus serotype 40 and 5 in the environmental samples. Overall, the results of this study suggest that real-time PCR technology permits the rapid detection of human viruses and indicator bacteria in the aquatic environment.

Benefits:

- Provides real-time PCR methods for detection of human adenoviruses and enterococci bacteria in environmental samples.
- Provides knowledge of diversity of human adenoviruses in the aquatic environment.
- Provides understanding of immunomagnetic separation and recovery of human adenoviruses.
- Provides information on a new tissue culture cell line for assaying the infectivity of human adenoviruses in environmental samples.
- Provides knowledge of the relationship between quantitative real-time PCR, adenovirus infectivity, and cultured enterococci bacteria in environmental waters and sewage.

Keywords: Real-time PCR, indicator bacteria, human viruses, *Enterococcus*, adenoviruses, coliphage, immunomagnetic separation, tissue culture, infectivity, sewage, coastal water.

TABLE OF CONTENTS

LIST OF TABLES

LIST OF FIGURES

LIST OF ACRONYMS

Ad	adenoviruses
PCR	polymerase chain reaction
RT-PCR	reverse transcription polymerase chain reaction
IMS	immunomagnetic separation
HEK-293A	human embryonic kidney cell 293A
VBNC	viable but noncultivable

EXECUTIVE SUMMARY

Waterborne pathogens are a significant cause of human disease world wide. To protect public health, the World Health Organization and the U.S. Environmental Protection Agency have established microbial pollution indicator standards and recommended routine monitoring of drinking and recreational waters for total and fecal coliforms and enterococci. However, current monitoring methods take 24 to 48 hours from the time the water sample is collected to the report of the assay results, which may result users being exposed to contaminated waters. Thus, a rapid monitoring method is urgently needed for human health protection.

Furthermore, bacterial indicators may not reflect the viral quality of the water because viruses are more resistant to sewage treatment and degradation in the environment. The absence or a low concentration of indicator organisms in water may not adequately reflect the absence of human viruses. In the authors' previous study of Southern California coastal waters, over 30% of coastal waters tested contained human adenoviruses, and the presence of these viruses did not correlate with an elevated level of bacterial indicators. The goal of this research was to develop and validate a molecular method for rapid and specific detection of microbial contaminants including human viruses and bacterial indicators in treated sewage effluents and receiving waters.

A review of molecular alternatives for indicator and viral pathogen detection was conducted. Although real-time PCR methods are widely used in clinical research for the detection of human viruses, environmental applications of this method are limited at present. In addition, efficient methods for concentration and purification of human viruses in complex environmental matrices are urgently needed. In this study, the authors designed, tested, and optimized real-time quantitative PCR methods using specific and degenerate primers and probes targeting adenoviruses and enterococci bacteria, respectively. Experimental testing of real-time PCR primers and probes for adenoviruses demonstrated reproducible results with an estimated efficiency of greater than 90% over a 5-log dynamic range of target concentration. The estimated efficiency of the enterococci bacteria real-time PCR assay was 99% over a 7-log dynamic range of target concentration. Application of these methods to sewage effluents and coastal waters demonstrated that real-time PCR methods are more sensitive than culture-based methods, suggesting a great potential for real-time quantification of microbial contaminants in recreational and drinking waters. However, real-time PCR methods, like other molecular detection technology, overestimate the concentration of infectious bacterial and viral concentration.

In addition to real-time PCR method development, the study also sampled and isolated human adenoviruses from Newport Bay, Calif. using human embryonic kidney cells, 293A. Cloning and sequencing of selective environmental adenovirus hexon gene suggested that most of the viruses recovered from the environment belong to adenovirus serotype 40. This result is in agreement with clinical data on the load of viral shedding in feces. Since adenovirus serotypes 40 and 41 are the major cause of childhood diarrhea, the result of this investigation indicates the importance of monitoring water for viral quality. Comparison of four different tissue culture cell lines for their sensitivities to adenovirus infection has demonstrated that the genetically modified 293A cells are the most efficient at recovery of adenoviruses 40. This result is a significant contribution to our ability to assay for infectious adenoviruses in environmental samples.

Immunomagnetic capture of human adenvoriuses was tested using paramagnetic beads coated with three different monoclonal antibodies and one multiclonal antibody targeting an adenovirus surface protein. However, neither antibody nor combination of the antibodies demonstrated sufficient specificity and efficiency at capturing human adenoviruses. Nonspecific binding with unlabeled magnetic beads was also observed for adenoviral particles.

Application of the real-time PCR method developed above has shown up to 10^4 adenovirus genome equivalence per liter of water in the Los Angeles and San Gabriel Rivers. However, none of the PCR-positive samples tested positive for infectious viruses using tissue cultures, suggesting that the real-time PCR method significantly overestimates the presence of infectious human viruses. Furthermore, although real-time PCR is capable of detecting adenoviral particles seeded in sediment samples, the amplification efficiency is greatly reduced due to the presence of PCR inhibitors in the sediment extracts. Sonication increases the release of viral particles attached to the sediment but also increases the concentration of PCR inhibitors released from the sediment. Further development of nucleic acid purification methodology will greatly enhance the ability to detect pathogens by real-time PCR in the environment.

In summary, the research presented here represents a significant advancement in the current ability to rapidly detect viruses and indicator bacteria in the environment. However, to assess public health risk, the application of this technology requires a better understanding of the relationship between genomic equivalence of target organisms and their infectivity.

CHAPTER 1.0

PROJECT CONCEPT

1.1 Rationale of Study

Microbial quality of water is a prime public health concern today. The risk of becoming ill as a result of microbial contamination of food and water is 100,000 times greater than the risk from pesticide contamination[1]. To protect public health, the World Health Organization and the U.S. Environmental Protection Agency (U.S. EPA) have established microbial pollution indicator standards and recommended routine monitoring of water for both total and fecal coliforms[1,3,4].

However, the adequacy of current water quality standards to indicate the presence or absence of human pathogens is still questionable. For example, human viruses are more resistant to sewage treatment processes and environmental conditions than bacterial indicators and therefore may pose a substantial threat. The absence or a low concentration of indicator organisms in water may not adequately reflect the absence of human viruses. In a previous study of Southern California coastal waters, the researchers found over 30% of coastal waters tested contained human viruses, and the presence of these viruses did not correlate with an elevated level of bacterial indicators[5]. Waterborne infectious viruses are suspected to be more important than current knowledge suggests because viral diseases are hard to identify by current diagnostic techniques. While the causative agent of nearly 50% of all acute gastrointestinal illness cannot be identified [6] most of these illnesses are suspected to be viral. Therefore, viral quality of recreational coastal water is of particular importance and a rising public health concern.

With rapid population growth and urbanization occurring along coastal regions around the world, the quality of coastal water and water supply is a global concern. Because Southern California is one of the most densely populated coastal regions in the U.S., it offers an opportunity to understand the intricate relationship between water quality and rapid human development in a coastal region. Population growth results in conversion of open land into nonpermeable surfaces, which increases the rate of runoff and affects coastal water quality through addition of sediment, toxic chemicals, microbial pathogens, and nutrients. In addition, a high population density also increases the volume of domestic sewage. Southern California coastal waters receive daily discharges from 15 municipal wastewater treatment facilities, eight power generating stations, 10 industrial treatment facilities, and 18 oil platforms. The Orange County Sanitation District (OCSD), the largest wastewater treatment facility in Orange County, Calif., discharging an average 250 million gallons per day of wastewater, is one of the few wastewater treatment facilities holding a national Ocean Discharge Permit to dispose of a mixture of primary and secondary treated sewage to the offshore Pacific Ocean. Furthermore, population growth is also coupled with increasing water consumption. To meet the increasing water demand of a growing Orange County, treated sewage water is used to replenish the groundwater basin.

The goal of this research was to use Orange County, Calif. as a test bed to develop and validate a method for rapid and specific detection of microbial contaminants of public health concern in treated sewage effluents and receiving waters. The method developed in this study will have a global application in understanding the removal of pathogens during sewage treatment processes and their potential transport and fate in offshore and coastal waters.

Currently, coastal recreational water quality in California is governed by standards set by the California Ocean Plan for total coliforms, fecal coliforms, and enterococci. Since the adoption of enterococci standards (California Assembly Bill 411) in 1999, the frequency of beach postings and closures has increased due to exceedance of enterococci standards. However, in many instances little is known regarding the source of contamination. In Huntington Beach, Calif., wildlife residing at the coastal salt marsh is suspected as the pollution source, but current methods cannot discriminate human sewage from animal fecal contamination. In addition, the current monitoring system requires at least 24-72 hours for bacterial culturing, which seriously delays the information transfer to the public and immediate beach management actions. For example, an accidental sewage spill in north Orange County in summer 2000 was not detected until a few days after the incident, resulting in exposure of bathers to sewage-contaminated water and hospitalization of several surfers.

Another important limitation of the current fecal pollution indicator system is its inability to indicate the viral quality of water. More than 100 different types of viruses are found in human waste and potentially transmitted by water[7]. These viruses are more resistant to environmental conditions and sewage treatment processes, including chlorination and UV radiation, than many of the sewage-associated bacteria[8]. In laboratory studies, enteric viruses have been reported to survive from 2-130 days in seawater. These survival periods surpass those reported for coliform bacteria in similar environments[9] and demonstrate a need to directly assay for human viral pathogen presence.

To fill the gaps in the current microbial indicator system, this research has developed a molecular method for rapid detection of human viruses and bacterial indicators in aquatic environments. This approach offers several advantages: 1) Viruses are species specific. Therefore, the presence of human viruses will indicate the presence of human fecal source contamination, either from a point source, such as leakage from sewage lines, or nonpoint source, such as urban runoff. 2) Real-time PCR analysis of viruses and indicators is quantitative and avoids time-consuming culturing and gel electrophoresis methods. The entire assay can be completed in a few hours, thereby increasing the speed of analysis. 3) Direct detection of specific human viruses will provide an accurate representation of the viral quality of recreational water.

1.2 Project Objectives

With the above goals in mind the objectives of this research were to:

- Develop and improve a real-time PCR method for rapid detection of human viruses and bacterial indicators in aquatic samples.
- Apply and validate this method for detection of target organisms in samples of treated sewage effluents, receiving coastal waters, and sediment in order to understand the efficacy of treatment processes and the source, transport, and fate of microbial contaminants in the environment.
- Compare the results from the real-time PCR method to the current bacterial indicators and the presence of F-specific coliphage in natural samples.

1.3 Project Background

1.3.1 Selection of Target Organisms

Viral infection in the U.S. may be the number one public health threat transmitted by water. Current estimates are that viral gastroenteritis produces 30%-40% of the cases of infectious diarrhea in the U. S., far outnumbering documented cases of bacterial and parasitic diarrhea. The cause of approximately 40% of presumed cases of infectious diarrhea remains unknown[10]. Over 100 different human viruses are associated with human waste, and all are potentially water transmissible. It is impractical, if not impossible, to detect all these viruses. However, certain human viruses may serve as an index for the presence of human fecal contamination and the viral quality of water. An ideal index virus should be the most prevalent virus in human waste, always present in the presence of other viruses, and at least equally resistant, if not more, as the rest of the human viruses to sewage treatment processes and environmental degradation conditions. Human adenovirus is a possible candidate to fill this role.

Adenovirus is a double-stranded DNA virus, belonging to the family adenoviridae. Most adenoviruses generally cause respiratory infections, but they can also be found in healthy persons in tonsils or adenoid tissue. Enteric adenoviruses are the only human enteric viruses to contain DNA rather than RNA. Specifically, adenoviruses 40 and 41 have been recognized as important etiological agents of gastroenteritis in children[11,12], second in importance only to rotaviruses. Respiratory adenoviruses are capable of causing large, persistent epidemics, and they may be more important than enteric viruses as a contagion during recreational water activities because of their potential transmission in aerosols[13]. Nonenteric adenoviruses have been associated with recreational water outbreaks of pharyngoconjunctival fever[14]. So far there has been no indicator of water-contact diseases other than enteric diseases.

Adenoviruses are frequently found in urban rivers[15-17], as well as polluted coastal waters[5,18,19]. Studies conducted in Europe have suggested using adenovirus as an index of human viral pollution since this virus was often detected in human fecal contaminated environmental samples[18]. All samples that were positive for enteroviruses or hepatitis A viruses also tested positive for human adenoviruses. Adenovirus infections occur year-round, and there is little or no seasonal variation in shedding[20]. They have been consistently found in similar or greater numbers than enteroviruses in raw sewage around the world[21-23] and are substantially more stable than either poliovirus or HAV in tap water and seawater[24]. For example, adenovirus survives three to five times longer in seawater than poliovirus, taking 99 days for a 2-log reduction of infectivity[24].

In spite of the above-mentioned potential, the utility of adenovirus as a viral contamination index is far from substantiated. The transmission of enteric adenovirus via water has never been well documented. Traditionally, the enterovirus family, including polio, coxackie, echo, and enteroviruses, has been used as a parameter for evaluating viral pollution of the environment because most of members can be isolated and quantified as PFU in cell culture[25]. Enteroviruses are the only viral parameter included in the guidelines of the European Community for water quality in the aquatic environment. The genetic sequencing of enteroviruses has identified extensive homology among family members' genomes. Based on this, pan-enterovirus probes and primers have been designed to detect the presence of any member of the family[26]. Some researchers have also suggested the use of poliovirus for viral monitoring because of its prevalence as a component of the human vaccine, but poliovirus is not always detected in wastewater[19]. The U.S. EPA describes the enteric virus group (including adeno, entero, hepatitis A, calici, rota, astro and coronaviruses) as the most reliable and effective virus index for environmental monitoring[27].

After reviewing the current state of knowledge, the researchers decided to use adenovirus as the target virus to confirm if any of the viruses could be used as a viral index. Further, comparing adenovirus with the current indicator organisms and other potential indicators could reveal the adequacy of current bacterial indicators. Another reason to select these viruses is that they are well studied, and genetic sequence information on them is readily available through the World Wide Web, making it feasible to design new primers to improve amplification specificity and efficiency.

Three types of fecal indicator bacteria are currently used as standards to indicate the quality of water: total coliform, fecal coliform, and *Enterococcus*. Among them, only *Enterococcus* was shown to directly relate to swimming-associated gastroenteritis at marine and freshwater bathing beaches[28,29]. Enterococci, formerly included within the fecal streptococci, were recently adopted by the State of California as an additional standard indicator for determining marine recreational water safety (State Assembly Bill 411) because this group of bacteria survive longer in marine environments and are more resistant to wastewater treatment processes than fecal coliforms[30,31]. Therefore, the researchers decided to use *Enterococcus* as the target bacteria indicator for real-time PCR development.

1.3.2 Concept and Background of Real-time PCR

Although real-time PCR should not be interpreted as real-time monitoring of pathogens in water, it does allow on-line monitoring of PCR product formation during PCR cycling, avoid post-PCR gel electrophoresis, and shorten pathogen detection time to a few hours.

All real-time PCR systems rely on the detection and quantification of a fluorescent reporter, the signal of which increases in direct proportion to the amount of PCR product in a reaction. In the simplest and most economical format, that reporter is the double-strand DNA-specific dye SYBR Green (Molecular Probes). SYBR Green binds double-stranded DNA and, upon excitation, emits light. Thus, as a PCR product accumulates, fluorescence increases.

The advantages of SYBR Green are that it's inexpensive, easy to use, and sensitive. The disadvantage is that SYBR Green will bind to any double-stranded DNA in the reaction, including primer-dimers and other nonspecific reaction products, which results in an overestimation of the target concentration. For single PCR product reactions with well-designed primers, SYBR Green can work extremely well, with spurious nonspecific background only showing up in very late cycles.

The two most popular alternatives to SYBR Green are TaqMan and molecular beacons, both of which are hybridization probes relying on fluorescence resonance energy transfer (FRET) for quantification.

TaqMan Probes (Applied Biosystems, Forest City, Calif.) are oligonucleotides that contain a fluorescent dye, typically on the 5' base, and a quenching dye, typically located on the 3' base. When irradiated, the excited fluorescent dye transfers energy to the nearby quenching dye molecule rather than fluorescing, resulting in a nonfluorescent substrate. TaqMan probes are designed to hybridize to an internal region of a PCR product. During PCR, when the polymerase replicates a template on which a TaqMan probe is bound, the 5' exonuclease activity of the polymerase cleaves the probe. This separates the fluorescent and quenching dyes and FRET no longer occurs. Fluorescence increases in each cycle, proportional to the rate of probe cleavage. Molecular beacons also contain fluorescent and quenching dyes, but FRET only occurs when the quenching dye is directly adjacent to the fluorescent dye. Molecular beacons are designed to adopt a hairpin structure while free in solution, bringing the fluorescent dye and quencher in close proximity. When a molecular beacon hybridizes to a target, the fluorescent dye and quencher are separated, FRET does not occur, and the fluorescent dye emits light upon irradiation. Unlike TaqMan probes, molecular beacons are designed to remain intact during the amplification reaction and must rebind to target in every cycle for signal measurement.

TaqMan probes and molecular beacons allow multiple DNA species to be measured in the same sample (multiplex PCR), since fluorescent dyes with different emission spectra may be attached to the different probes. Multiplex PCR allows internal controls to be co-amplified and permits allele discrimination in single-tube, homogeneous assays. These hybridization probes afford a level of discrimination impossible to obtain with SYBR Green because they will only hybridize to true targets in a PCR and not to primer-dimers or other spurious products.

Specific requirements should be followed in designing TaqMan primers and probes to maximize the sensitivity of the detection system.

TaqMan Probe:

- Keep the G-C content in the 30-80% range.
- Avoid runs of an identical nucleotide, especially for guanine, where runs of four or more Gs should be avoided.
- Do not put Gs on the 5' end.
- Select the strand that gives the probe more Cs than Gs.
- For single-probe assays, Tm should be 68-70°C when using Primer Express software.
- Length between 20 and 30 nt and a Tm of 10°C higher than those of primers.

Primers:

- Choose the primers after the probe.
- Design the primers as close as possible to the probe without overlapping the probe.
- Keep the G-C content in the 30-80% range.
- Avoid runs of an identical nucleotide, especially for guanine, where runs or four or more Gs should be avoided.
- When using Primer Express software, the Tm should be 58-60°C.
- The five nucleotides at the 3' end should have no more than two G and/or C bases.

CHAPTER 2.0

REAL-TIME PCR DETECTION OF HUMAN ENTERIC ADENOVIRUS SEROTYPE 40

2.1 Introduction

Human enteric viruses represent a diverse group. The term enteric virus reflects the fact that most of the species inhabit the alimentary (enteric) tract. Most of them are nonenveloped DNA viruses. The Adenoviridae family is the only enteric virus that contains the double-stranded DNA genome[32]. Table 2-1 lists the major viral species belonging to the enteric viruses group.

Table 2-1. Human Enteric Viruses.

Family	Genus	Species	Genome
Adenoviridae	Mastadenovirus	Human adenovirus[1] serotypes 40 and 41	Linear dsDNA
Picornaviridae	Enteroviruses	Polio 1, 2, 3, Coxsackie[2] A 1-24 & Coxsackie B 1-6, Echo[3] 1-34, Entero 68-71	Linear positive-sense ssRNA
	Hepatovirus	Hepatitis A (Entero 72)	
Astroviridae	Asterovirus	Asterovirus	Positive sense – ssRNA
Reoviridae	Rotavirus	Rotaviruses (Group B)	10 to 12 segments linear dsRNA
Caliciviridae	Calicivirus	Calicivirus; Norovirus	Positive-sense-ssRNA

[1]Include a total of 51 serotypes but only 2, 5, 40 and 41 cause gastroenteritis.
[2]Named for the hometown of the first patient, Coxsackie, NY.
[3]Enteric cytopathic human orphan.

The host enteric viruses must survive the acidic environment of the stomach and the bile salts of the upper small intestine, penetrate the mucus coating of the gut, and bind to the proper receptors. A vigorous humeral response that can neutralize the virus before it initiates infection must also be circumvented. Symptoms range from gastroenteritis to more life-threatening diseases such as myocarditis and aseptic meningitis.

Human adenoviruses have been recovered from virtually every human organ system and have been associated with a wide spectrum of clinical disease[33]. There are presently 49 recognized adenovirus serotypes defined by neutralization with type-specific animal antisera, and candidate serotypes 50 and 51 have recently been reported. Adenovirus serotypes can be classified into six species (formerly subgenera), designated species A through F. This classification scheme is generally consistent with subgroupings of adenoviruses on the basis of their physio-chemical, biological, and genetic properties and is diagnostically important in that infections with different adenovirus species and serotypes are often associated with distinct clinical outcomes and epidemiologic features[34]. Certain types of adenovirus are commonly associated with particular clinical syndromes are respiratory illness, pharyngitis, gastroenteritis, conjunctivitis, keratoconjunctivitis, pneumonia, acute hemorrhagic cystitis, and hepatitis[35]. Gastroenteritis is most commonly associated with serotypes 40 and 41, but others may be involved. Adenovirus 40 (Ad40) and 41 (Ad41) together account for approximately 50% of all types of adenoviruses in stool specimens[18]. They are also recognized as one of the major causes of childhood diarrhea.

Recently, adenoviruses have been placed on the U.S. EPA drinking water Contamination Candidate List (CCL) because they are more resistant to traditional UV disinfection than fecal indicator bacteria during water treatment[36]. The number one priority for effective monitoring of microbial contaminants on CCL is to identify a method that is capable of rapid, sensitive, and accurate detection of the load of microbes. So far, no such method exists for rapid and sensitive detection of the load of Ad40 in environmental waters. Thus, the objective of this study was to develop a real-time PCR method for quantification and detection of human enteric adenovirus serotype 40 for application in complex environmental matrices. Real-time PCR methods built on the specific detection of adenovirus serotype 40 hexon gene is reported in this chapter.

2.2 Materials and Methods

2.2.1 Primers and Probe Selection

To identify the conservative sequence among adenovirus genomes, DNA sequence of adenovirus serotype 40 (Ad40) hexon gene was submitted for a BLAST search to the NCBI database. A 251 bp region within the hexon gene was identified to best match to 28 different serotypes of human adenoviruses. In addition, significant sequence homologies were also observed with adenoviruses of feline, simian type 25, bovine, canine type 1 and 2, porcine type 3 and 5, and equine type 1 and 2 within this region. Using Primer Express Software version 2.0 (Applied Biosystems), a probe was first selected following the manufacturer's instruction (see Chapter 1.0). Primers adjacent to the probe were then designed using the same software. Both primers and probe sequences are listed in Table 2-2. All primers and probes for this study were purchased from Applied Biosystems (Forest City, CA).

Table 2-2. Primers and Probe for Detection of Human Adenovirus 40 Using Real-time PCR and Construction of Internal Positive Control (IPC).

Target	Primers	Probe	Accession
ADV40	AD1 (F): ACCCACGATGTAACCACAGACA AD2 (R): ACTTTGTAAGAGTAGGCGGTTTCC Product size: 88 bp (157-245)	6-FAM- CTGCGCTTCGTGCCCGTCG-TAMRA (196-214)	ADV Hexon X51782
IPC[1] ADV40	F: CCCCCTCGATGATGCCGCAA R: CTGTCCACTGCCTGGTTCCA Product size: 1090 bp (8-1097)	N/A	X51782

[1]Internal Positive Control
6-FAM; 6-carboxyfluorescein
TAMRA; 6-carboxytetramethyl rhodamine.

2.2.2 Construction of Real-time PCR Internal Standard

To achieve absolute quantification requires an internal standard with known copy number of target to establish a quantification standard curve. A 1090 bp hexon region of Ad40 was amplified using primers specific to Ad40 (Table 2-2). The amplicon was subsequently ligated into the pGEM-T Vector (Promega), transferred into *E.coli* JM109 competent cell and platted on LB agar containing ampicillin, IPTG (isopropyl-beta-D-thiogalactopyranoside), and X-Gal (5-bromo-4-chloro-3-indolyl-beta-D-galactopyranoside) following manufacturer's instructions. White colonies were selected and analyzed for the presence of insert within the plasmid. Using the recombinant plasmid as templates, the amplification of 88 bp real-time PCR target was confirmed with conventional PCR and with real-time PCR (Figure 2-1). The plasmid with insert is designated pAD40.

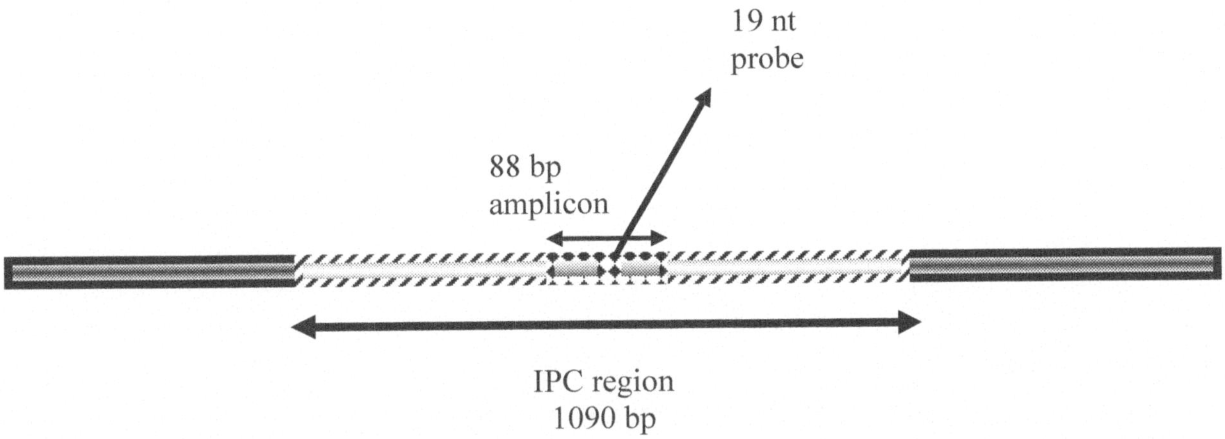

Figure 2-1. Schematic Representation of pAD40 (1091 bp).

2.2.3 Real-time PCR with SYBR Green

To test the efficiency of amplification using real-time PCR primers designed in this study, SYBR Green dye was first used because this method provides a relatively simple and inexpensive way for empirically testing the real-time PCR amplification. Real-time PCR with SYBR Green was performed in an ABI Prism 7000 sequence detection system (Applied Biosystems). The reactions were carried out in a 96-well plate in a 25μl reaction volume containing 12.5 µl of 2× SYBR Green master mix (Applied Biosystems), a 900 nM concentration of each forward and reverse primer, dNTP mix with dUTP, AmpliTaq Gold and 1 ng of plasmid pAD40 DNA. The use of the hot-start enzyme AmpliTaq Gold DNA Polymerase in all SYBR Green PCR master mix yielded the highest performance available by minimizing nonspecific product formation. Using dNTP mix with dUTP (by replacing dTTP) allows PCR products to be degraded by UNG (Uracil-N-Glycosylase). Incubation of reactions with UNG prior PCR destroys any carryover dUTP-containing PCR products from previous reactions. This procedure is routinely used for carryover contamination control. A passive internal reference dye (ROX) was also included in the PCR mix to minimize well-to-well variability that can result from pippetting errors and sample evaporation, etc. To determine the PCR detection limits, the amounts of DNA used per reaction varied from 1.0 ng to 1.0 ag (10^7 to 10 copies of genome). The thermal profile for all PCRs used was 95°C for 10 min, followed by 40 cycles of 95°C for 15s and 60°C for 1min. Each sample was replicated two to three times. All reactions were repeated at least three times independently to ensure the reproducibility of the results.

2.2.4 Dissociation Curve Analysis

A dissociation curve is a graph that displays dissociation data from the amplicons of PCR runs. Dissociation curve analysis was performed after each real-time PCR assay using SYBR Green dye. Data was obtained by slowly ramping the temperature of reaction solutions from 60° to 95°C while continuously collecting fluorescence data. In addition, the absence of nonspecific amplification was confirmed by analyzing the PCR amplification products by traditional agarose gel electrophoresis to assist with interpretation of the dissociation curve.

2.2.5 Quantification of Adenovirus Hexon Gene by TaqMan PCR

Amplification was performed in a 25 µl reaction mixture with a TaqMan PCR core reagent (Applied Biosystems). Each reaction mixture contained 1 µl of Ad40 viral frozenstock (concentration not determined) or 1 ng of plasmid pAD40 purified by the QIAgen miniprep kit (Qiagen); 2x TaqMan core buffer; 5 mM $MgCl_2$; a 200 µM concentration (each) of dATP, dCTP, and dGTP; 400 µM dUTP; a 900 nM concentration of each primer; 250 nM probe (see below for primers and probe optimization); 0.25 U of AmpErase uracil *N*-glycosylase; and 1 U of AmpliTaq Gold polymerase. Following activation of the uracil N-glycosylase (2 min, 50°C) and activation of the AmpliTaq Gold for 10 min at 95°C, 40 cycles (15 s at 95°C and 1 min at 60°C) were performed with an ABI 7000 sequence detector system (Applied Biosystems). Each run contained at least three no-template negative controls to establish the baseline emission intensity of the quenched reporter dye.

2.2.6 Optimizing Primers and Probe Concentrations

The purpose of this experiment was to determine the optimal primer concentrations that yield the lowest threshold cycle (Ct) and maximum ΔR_n while minimizing nonspecific amplification. By independently varying forward and reverse primer concentrations, the concentrations that provide optimal assay performance can be identified (spanning an initial concentration range of 50 nM to 900 nM for each primer). The primer concentrations used in the primer optimization matrix are shown in Table 2-3.

Table 2-3. Primer Optimization Matrix.

Reverse Primer (nM)	Forward Primer (nM)		
	50	300	900
50	50/50	300/50	900/50
300	50/300	300/300	900/300
900	50/900	300/900	900/900

Using the optimal primer concentrations defined by the primer optimization matrix ensures assay performance when using a 250 nM probe concentration. However, depending on the requirements of the assay, a probe optimization experiment in which the probe concentration is varied from 50–250 nM can prove useful. However, to ensure the best reproducibility, especially when detecting low copy numbers of a target sequence, probe-limiting concentrations should be avoided. By using a 250-nM concentration, probe limitation is avoided and large ΔR_n values are ensured.

2.2.7 Real-time Detection of Adenovirus Hexon Gene in Environmental Samples

Three sets of experiments were conducted to test the application of real-time PCR in environmental samples. Secondary sewage from a local sanitation district was used as the environmental matrix for testing. For the first set of experiments, serial dilutions of plasmid pAD40 (1 to 100 pg) were seeded to 1 ml of secondary sewage. Sterile water was seeded in parallel to serve as a control matrix for the seeding experiment. In the second set of experiments, *E. coli* cells containing plasmid pAD40 were seeded to secondary sewage and sterile water at a final concentration of 10^2, 10^4, and 10^6 CFU/ml. In the third set of experiments, 1 µl of Ad40 viral frozen stock was seeded in 1 ml of secondary sewage in triplicate. Seeded water samples were either used directly for real-time PCR or further extracted/purified to remove PCR inhibitors. Nucleic acid extraction was performed according to the method described by Boom et al. [37] with minor modifications. In brief, 50 µl of sample was lysed by 900 µl of guanidinium thiocyanate (GuSCN) lysis buffer at room temperature for 10 min. Then 40 µl of silica particles (Sigma Chemical Inc.) was added and nucleic acid adsorbed at room temperature for 10 min with gentle shaking. Silica beads were pelleted, washed, and briefly dried. The nucleic acid was eluted from the beads using 50µl of Tris-EDTA buffer at a temperature of 56°C.

2.2.8 Virus Seeding Study in Various Environmental Matrices

Ad40, freshly harvested from tissue culture of human embryonic kidney cell 293A, was purified and used in the seeding study. A final concentration of 10^6 PFU/ml was seeded into each of the following environmental matrices: creek water, estuarine water, ocean water, and secondary sewage effluent. Sterilized DI water was used as the reference control matrix to determine PCR efficiencies in each type of environmental matrix. All environmental samples were collected in June 2004, using sterilized containers. Creek water was taken from San Diego Creek, a tributary of Newport Bay, CA. Estuarine water was taken from the middle of Newport Bay, ocean water was taken from Huntington Beach, CA, and the secondary sewage effluent was obtained from Orange Country Sanitation District, CA.

Viral nucleic acid was extracted from seeded samples using QIAmp Viral RNA Mini Kit (Qiagen) following manufacturer's recommendations. One microliter of each nucleic acid extract was used in the TaqMan real-time PCR reaction, using optimized condition described above.

2.3 Results

2.3.1 Evaluation of Primers and Probe

The specificity of the primers and probe was evaluated against the currently available sequence database using the BLAST program from NCBI (http://www.ncbi.nlm.nih.gov/) (Tables 2-4, 2-5, 2-6). The forward primer only had homology with human adenovirus serotypes 40 and 41 and the reverse primer was specific for human adenovirus serotypes 40, 3, 7, and 16 (Tables 2-4 and 2-5). Search results indicate that the probe was specific for 17 different serotypes of human adenovirus (Table 2-6). In addition, the probe had a significant homology with Simian adenovirus type 25. However, because the specificity of primer sequences used, these primers and probe set can only specifically detect human adenovirus serotype 40 but not any other serotypes.

Table 2-4. Results of Homology Search for Forward Primer.

ACCESSION	DEFINITION
L19443, X51782	Human adenovirus type 40, complete genome
D13781, X51783	Adenovirus type 41 hexon gene, complete cds

Note: Forward primer starts from nucleotide number 157 on hexon gene of Ad40.

Table 2-5. Results of Homology Search for Reverse Primer.

ACCESSION	DEFINITION
X76549, AJ250780	Adenovirus type 3 hexon gene
X76551, AF065065, AJ250781, Z48571	Adenovirus type 7 hexon gene
X74662, AJ250784	H. sapiens adenovirus type 16 hexon gene
L19443, X51782	Human adenovirus type 40, complete genome

Note: Reverse primer starts from nucleotide number 245 on hexon gene of ADV 40.

Table 2-6. Results of Homology Search for TaqMan Probe.

ACCESSION	DEFINITION
X76549, AJ250780	Adenovirus type 3 hexon gene
AF065064	Human adenovirus type 4 hexon gene, complete cds
AF515814, AF321311, AF065067, AF065066, AF065068, AF053087, AF053086, AF053085	Human adenovirus type 7, type 7a hexon protein gene, complete cds
AF161562	Human adenovirus type 9 hexon protein gene, partial cds
AF161563	Human adenovirus type 10 hexon protein gene, partial cds
AF161570, AJ250782, AB018424	Human adenovirus type 11 hexon protein gene, partial cds
AF161564	Human adenovirus type 13 hexon protein gene, partial cds
AF161571, AJ250783	Human adenovirus type 14 hexon protein gene, partial cds
X74662, AJ250784	H. sapiens adenovirus type 16 hexon gene
AF161565, AF161566	Human adenovirus type 19, type 19a hexon protein gene, partial cds
AB052911, AF161573, AJ250786, AB018426	Human adenovirus type 34 gene for hexon protein, complete cds
AF161574, AJ250787, AB018427, AB052912	Human adenovirus type 35 hexon protein gene, partial cds
X51782, L19443	Adenovirus type 40 hexon gene
AF394196	Simian adenovirus 25, complete genome
AF161568	Human adenovirus type 44 hexon protein gene, partial cds
U20821	Human adenovirus type 48 hexon capsid protein gene, partial cds

Note: The probe covers nucleotides 196 to 214 on hexon gene of ADV 40.

2.3.2 Real-time PCR with SYBR Green

Real-time PCR using the above primers and SYBR Green dye yielded a steady increase of fluorescence signals in samples containing Ad40 in comparison with background and no-template controls. This increase of fluorescence signal was also proportional to the initial concentration of the target, suggesting amplification of the desired target (data not shown). To further confirm this result, a dissociation curve analysis was run following each SYBR Green PCR assay. Figure 2-2 shows an example of a dissociation curve of SYBR Green PCR of Ad40 virus. Change in fluorescence was plotted against the change of temperature. The change in fluorescence is due to SYBR Green dye interacting with dsDNA. The increase in temperature causes PCR products to undergo denaturizing, a process accompanied by a decrease in fluorescence for solutions containing SYBR Green dye. By plotting the first derivative of fluorescence versus temperature, the melting profile of distinct PCR products is obtained, showing that the melting temperature of the product generated in the absence of a template is lower than the melting temperature of the specific product generated with template.

For all the SYBR Green PCR amplification described in this chapter, samples were routinely analyzed by their amplification profile and the dissociation curve to ensure the specificity of the intended product. When Ad40 and pAD40 samples were tested by SYBR Green PCR using the virus-specific primers designed above, a dissociation curve with a single sharp peak was obtained (Figure 2-2). A single dissociation peak with a Tm within a variation of 2°C of the expected temperature suggests an amplification of a specific product[38]. Since only a single peak was observed, the result indicated minimal primer-dimer formation during the PCR reaction. Therefore, this study indicates that the primers and condition used are optimal for assay of Ad40.

2.3.3 Real-time PCR with TaqMan Probe

Amplification of Ad40 using the above primers and FAM-TAMRA labeled TaqMan internal probe showed proportional increase of FAM signal above background value. The fluorescence changes during PCR cycles were recorded by the fluorescence component graph (Figure 2-3), which showed simultaneous increase of fluorescence signal of FAM dye and decrease of TAMRA dye. The graph of reference dye (ROX) is constant, which suggests amplification of the intended target is associated with excitation of FAM.

The increase of FAM relative to reference dye ROX is defined as ΔR_n. The PCR cycle at which a statistically significant increase in the ΔR_n is first detected is called the threshold cycle (Ct). Target DNA copy number and Ct values are inversely related, so Ct value is used as a measure of the sensitivity and efficiency of PCR reaction. Using primer matrix, the optimum concentration of reverse and forward primers, which yielded the lowest Ct value, was calculated to be 300/900 nM (Figure 2-4). However, because of the high specificity of primers, the maximum concentrations (900/900) gave very close Ct values to the optimum concentrations.

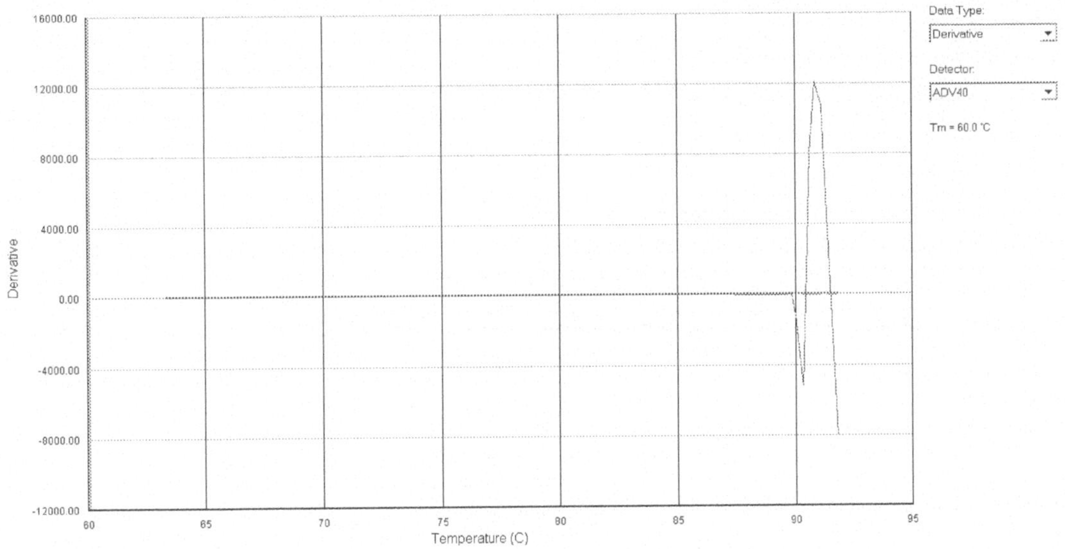

Figure 2-2. Dissociation Curve of SYBR Green Real-time PCR of Adenovirus Hexon Gene Showing Single Peak.

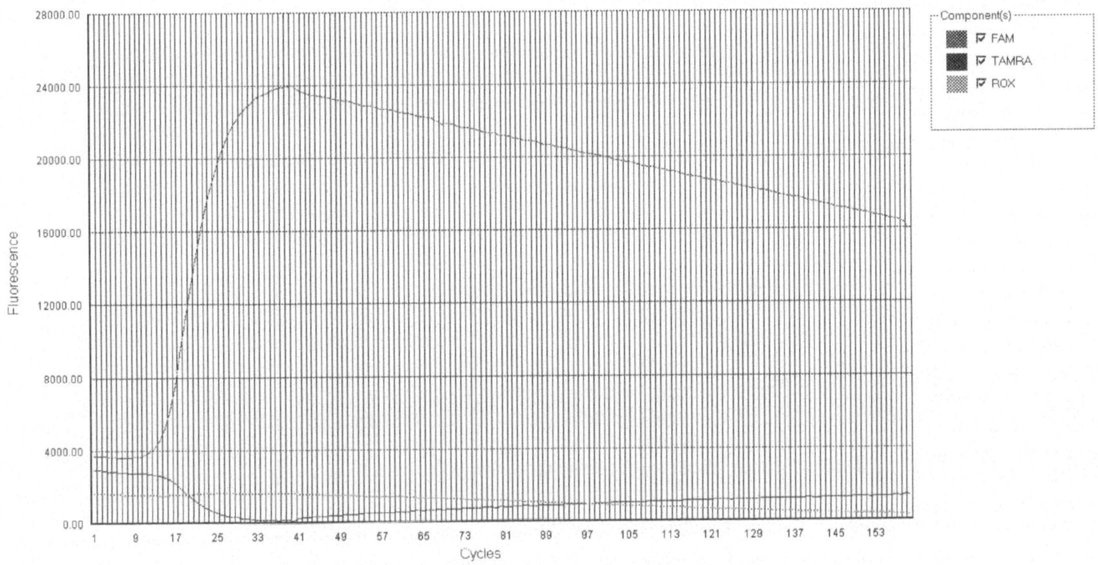

Figure 2-3. Fluorescence Component Graph of TaqMan PCR with pAD40.

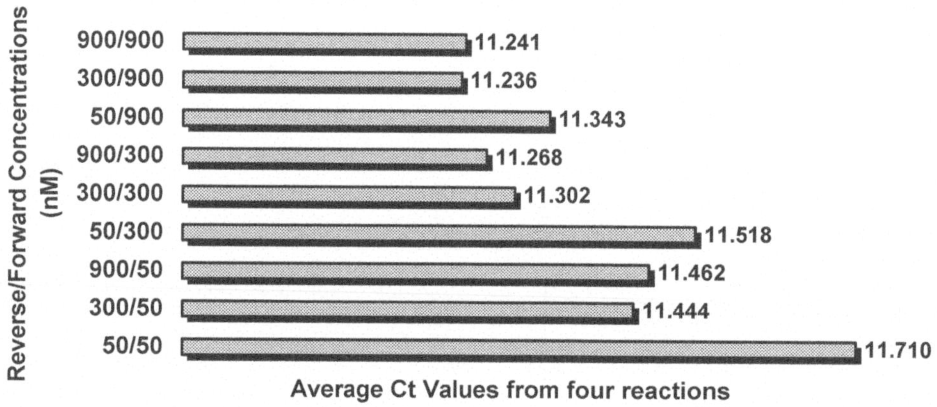

Figure 2-4. Optimization of Primer Concentrations.

2.3.4 Real-time PCR Quantification of Adenovirus Hexon Gene

Figure 2-5 shows the effect of initial DNA concentration Ct value. A tenfold dilution of target DNA resulted in ~3-cycle increase of Ct for TaqMan PCR of hexon gene. This relationship was also seen between the cycle number and relative fluorescence curve (Figure 2-6). A linear relationship between input DNA and Ct values over a range from 1 ng down to 100 ag was detected by TaqMan PCR. Detection of DNA concentrations over such a wide dynamic range is very useful for measuring viral loads.

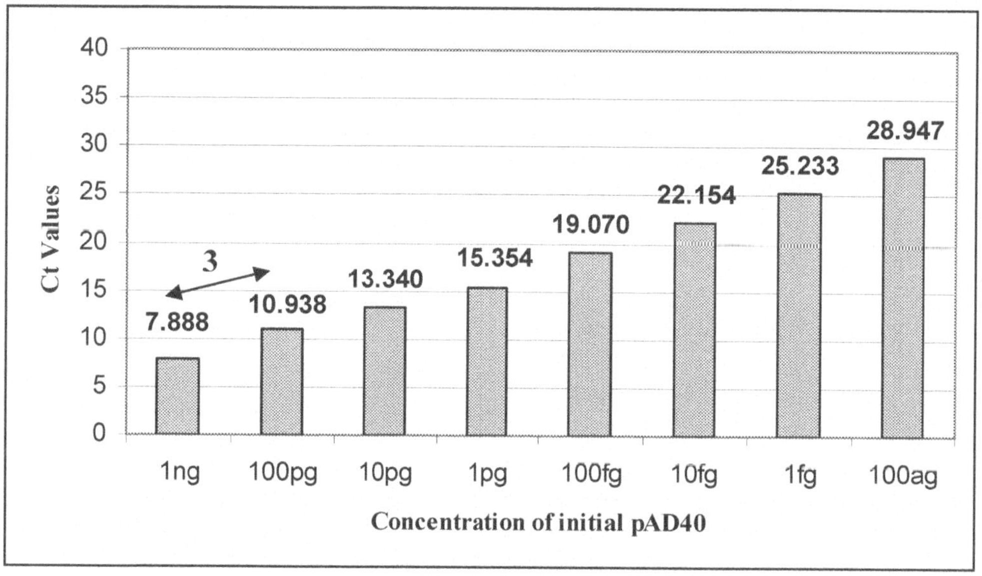

Figure 2-5. Effect of Initial DNA Concentration on Ct Value.

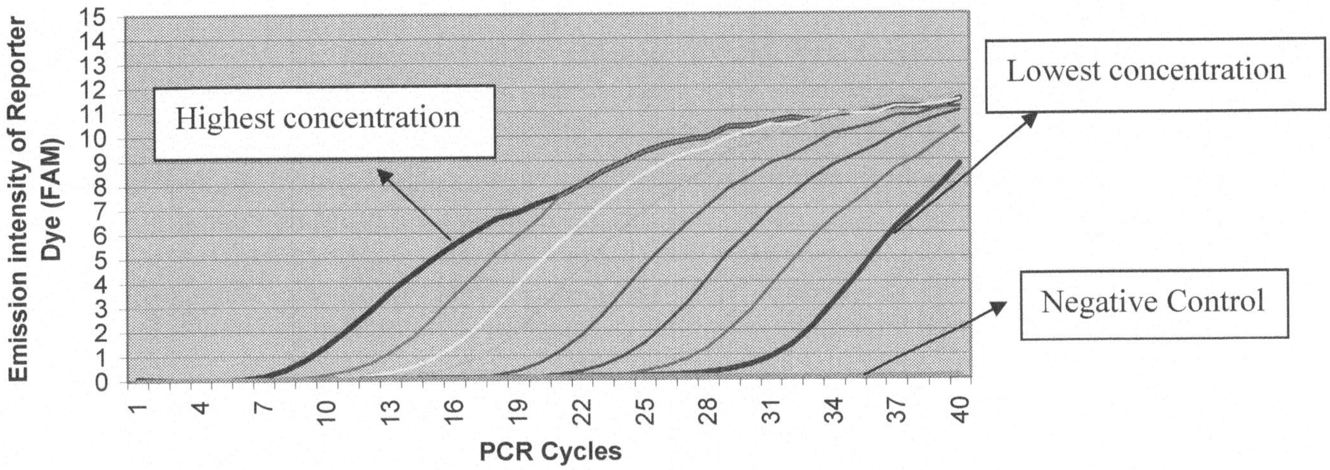

Figure 2-6. Effect of Initial DNA Concentration on Emission Intensity of Reporter Dye (FAM).
The highest concentration was 1 ng that gave the first signal around cycle number 7, and the lowest was 100 ng which gave the last signal around cycle number 28.

2.3.5 Standard Curve

Using a different concentration of plasmid pAD40, a standard curve was created (Figure 2-7). Amplification efficiency (E) was estimated by using the slope of the standard curve and the formula $E = (10^{-1/\text{slope}} - 1) \times 100$. A reaction with 100% efficiency generates a slope of -3.32. The efficiency of each assay was calculated from the slope of the standard curve. The standard curve generated from the PCR data resulted in a reaction efficiency of >95% for the pAD40, with correlation coefficients of 0.996982 and the slope of -3.194287 (Figure 2-7).

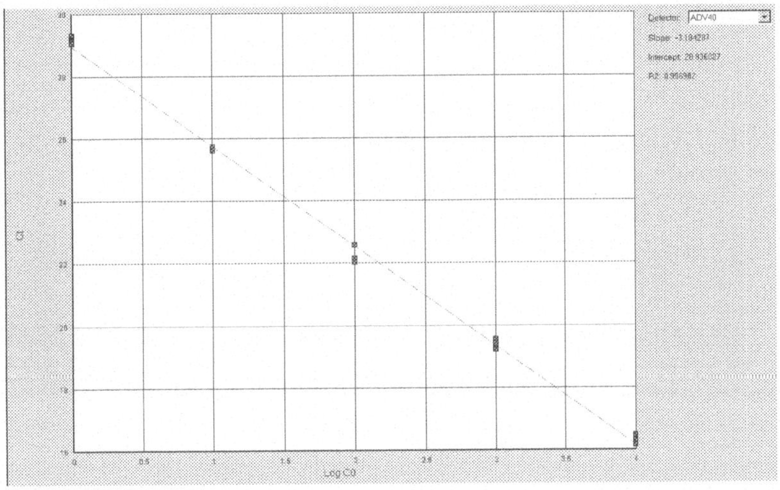

Figure 2-7. TaqMan PCR Standard Curve for pAD40.

2.3.6 Quantification of Ad40 Hexon Gene in Seeded Environmental Samples

Comparison of Ct values for different concentrations of pAD40 seeded in sterile water versus sewage showed that the Ct values were in general one to two cycles higher in the secondary sewage (Figures 2-8A and 2-8B). This result could be due to the presence of PCR inhibitors in the secondary sewage that result in a decrease of PCR efficiency. An alternative explanation may be that the presence of enzyme activity in the sewage caused degradation of naked plasmid DNA. Since the differences between seeded secondary sewage and sterile water were relatively small, the results indicate that the real-time PCR method is capable of applying to secondary sewage for viral load determination.

Little difference in Ct values was found between secondary sewage and sterile water seeded with *E.coli* with pAD40, except the samples contain the lowest concentration of cells (Figure 2-9). Interestingly the Ct values were lower in seeded sewage containing 10^2/ml *E. coli*, which may be due to the presence of background adenoviruses in the secondary sewage that contributed to the initial target concentration. Because this background signal was low in comparison with seeded target at high concentration, it was only observed at low seeding concentration. Similar results were also found when comparing *E. coli* seeded sewage and sterile water followed by nucleic acid extraction (Figure 2-10). At the lowest seeding concentration, the sewage sample yield a Ct value of 32.6 in comparison to the water sample of 37.1. However, no amplification signal was detected in the unseeded secondary sewage sample. Additional experiments should be conducted to verify the background level of adenovirus in secondary sewage (see Chapter 3.0 and 7.0). When Ct values of extracted/purified samples were compared with unextracted samples, no significant differences were observed, suggesting the PCR inhibition presented in the secondary sewage does not significantly reduce the efficiency of PCR reaction or the inhibition effect was not removed by nucleic acid extraction/purification process. Triplicate of Ad40 viral particle seeded sewage sample yielded very similar Ct values (Figure 2-11), suggesting the assay is reproducible within a single sample in the environmental matrix and the within-sample variability is at acceptable level.

To quantitatively compare the efficiency of real-time PCR in different environmental matrices, assays were performed using environmental matrices seeded with known titers of Ad40 particles. Table 2-7 shows that the method is applicable to quantify the Ad40 viral load in all environmental matrices tested. The viral load varied within one log unit (threefold differences between lowest and highest value) in different matrices. However, the efficiency of PCR reduced with the increased concentration of complex organic material in the sample. Secondary sewage effluent had the lowest PCR efficiency of 32%, while ocean water had relatively higher PCR efficiency due to the lower concentration of PCR inhibitors in the sample.

A) **Plasmid pAD40 in sterile water.**

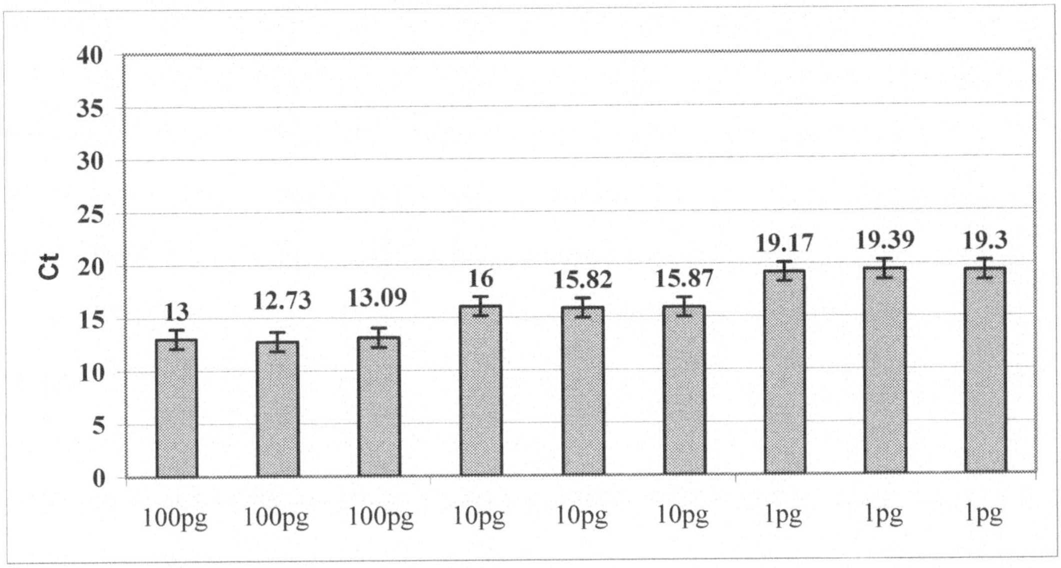

B) **Plasmid pAD40 in secondary sewage.**

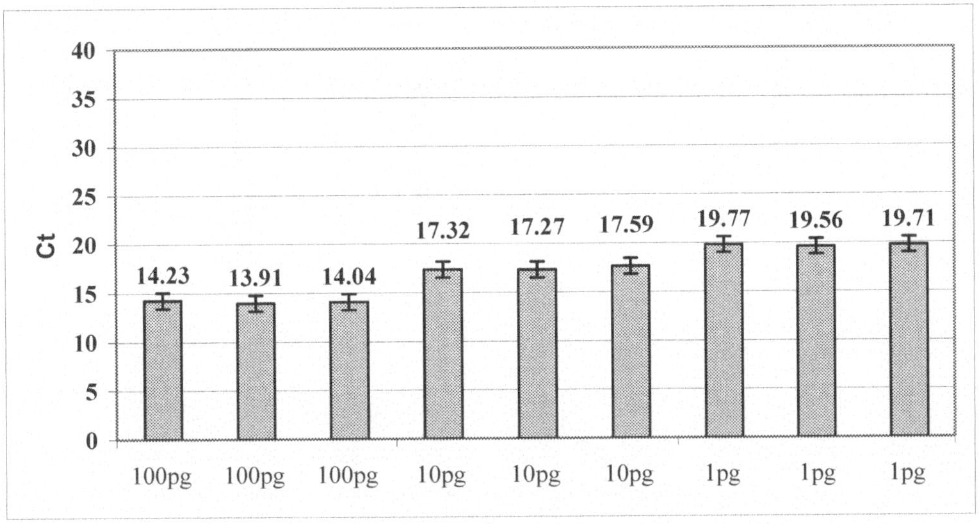

Figure 2-8. Comparison of Ct Values for Different Concentrations of pAD40 in Sterile Water Versus Sewage.

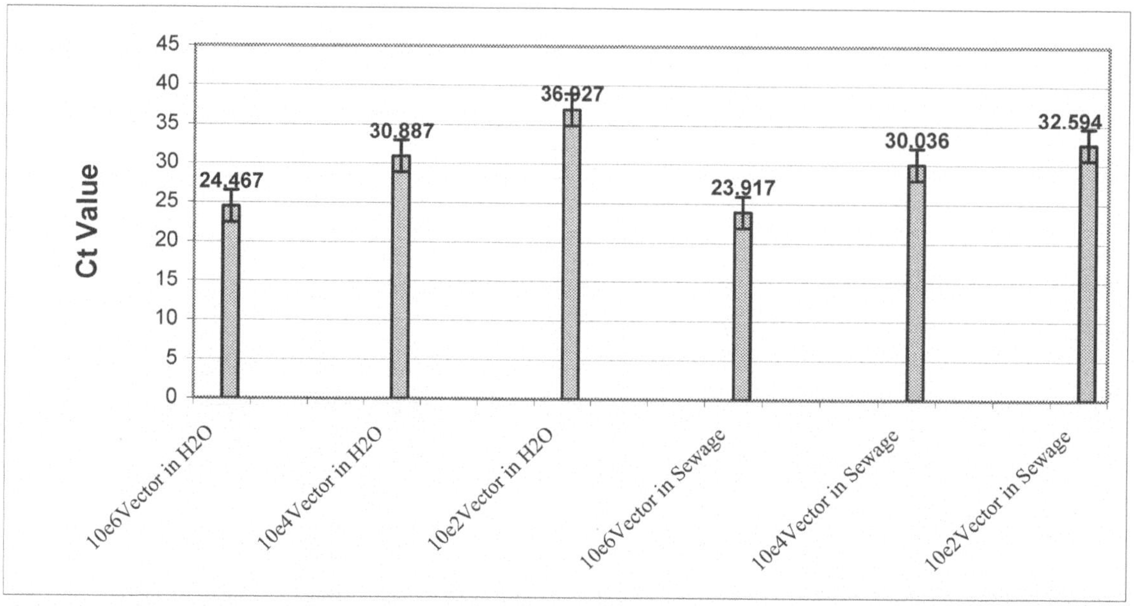

Figure 2-9. Comparison of Ct Values of Different Concentrations of Transformed *E.coli* with pAD40 in Water Versus Sewage.

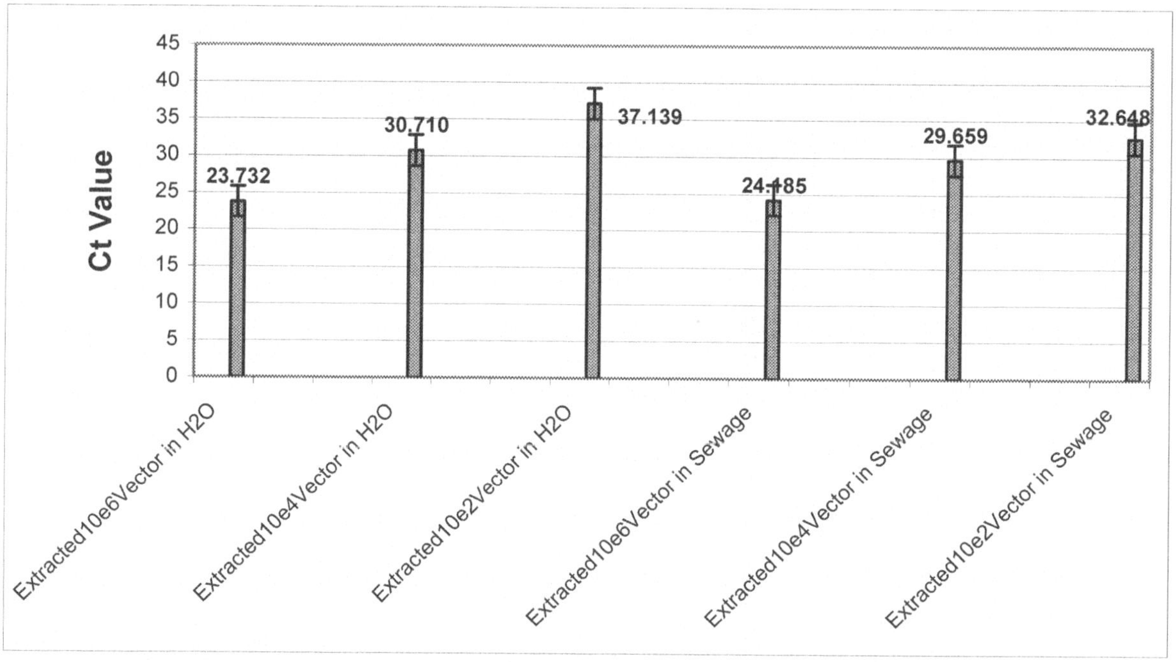

Figure 2-10. Comparison of Ct Values of Extracted pAD40 from Vector in Water Versus Sewage.

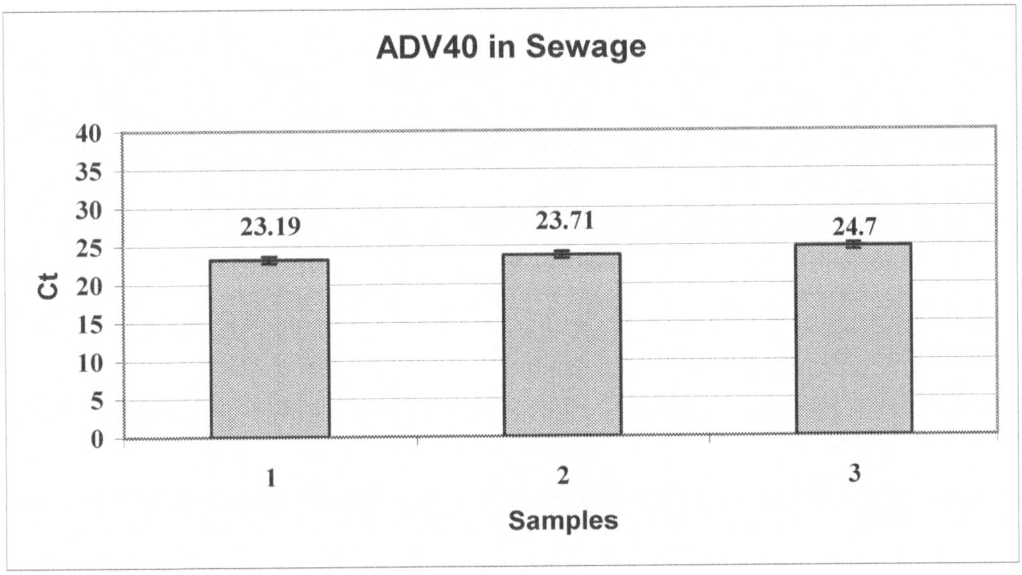

Figure 2-11. Reproducibility of TaqMan PCR Quantification of Adenovirus Sterotype 40 in Secondary Sewage Samples.

Table 2-7. Efficiency of TaqMan Real-time PCR in Different Environmental Matrices Seeded with Adenovirus Serotype 40.

Sample Matrices	Salinity (ppt)	Genome # by Real-time PCR	Real-time PCR efficiency
Sterile DI Water	0	1.6×10^6	100%
Secondary Sewage	0	5.1×10^5	32%
San Diego Creek Water	0.5	6.7×10^5	42%
Newport Bay Water	30	9.4×10^5	59%
Huntington Beach Water	34	9.7×10^5	61%

2.4 Discussion and Conclusions

Nucleic acid technology based on PCR has shown an increasing role in detection of viruses and bacteria in clinical and environmental laboratories. Real-time PCR techniques, involving fluorescent dyes or fluorophores with a spectrofluorometric thermal cycler, have been used for rapid quantification and identification of bacteria and viruses, including herpes simplex virus[39], varicella-zoster virus[39], cytomegalovirus[40], *Legionella pneumophila*[41], enterohemorrhagic *Escherichia coli*[42], *Staphylococcus aureus*[43], *Mycobacterium bovis*[44], and also for genes conferring drug resistance in *Helicobacter pylori*[45], hepatitis B virus[46], *S. aureus*[43] and *M. tuberculosis*[47].

Since the SYBR Green dye is more economical than TaqMan probes, the use of SYBR Green dye is very useful for primer evaluation studies before purchasing the probe. The optimal amplicon length for real-time PCR is less than 100 bp. In this study the researchers designed an amplicon of 88 bp; however, other investigators have used the TaqMan system to detect amplicons as long as 400 bp[48]. Shorter amplicons amplify more efficiently than longer ones and are more tolerant of reaction conditions because they are more efficiently denatured during the 92-95°C step of the PCR, allowing the probes and primers to compete more effectively for binding to their complementary targets. Because the extension rate of Taq polymerase is between 30 and 70 bases/second[48], polymerization times as short as 15s are sufficient to replicate the amplicon, making amplification of genomic DNA contaminants less likely and reducing the time it takes to complete the assay.

Primers are generally used in the 50-900 nM range according to manufacturer's recommendation (Applied Biosystems). Empiric testing, however, is necessary for optimization for each specific assay. The researchers found that primer concentrations of 300 nM and 900 nM are optimal for forward and reverse primer concentration, respectively, for this particular assay. Too great a primer concentration may promote miss-priming and accumulation of nonspecific product. Selecting primers that have only one or two G/Cs within the last five nucleotides at the 3' end can minimize nonspecific priming. This relative instability at their 3' ends makes them less likely to hybridize transiently and available for nonspecific extension by the DNA polymerize.

The optimal concentration of fluorigenic probes depends on background fluorescence in addition to primer concentration. Probe concentration is usually in the 250 nM range as recommended by the manufacturer (Applied Biosystems), and the probes must compete for binding to their target with the amplification primers, which are present in a greater concentration. Therefore, the Tm of the probes must be significantly greater (approximately 10°C) than that of the primers to ensure that they hybridize before the primers. An additional requirement of the hybridization probe system (TaqMan) is that the target sequence should be located towards the 3' end of the amplicon, close to the reverse primer-binding site on the opposite strand. As fluorescence is measured during the annealing phase, this arrangement allows time for the measurement to be made before the probe is displaced by the polymerase.

This stringent requirement for probe and primer design limits the selection of primers and probe. Although the primary goal of this study was to select primers and probe capable of detecting the maximum number of serotypes of human adenovirus without cross-reaction with nonhuman viral species, only human adenovirus serotype 40 is currently detectable using the primers and probe designed in this study. Therefore, in the next step of the study the researchers considered straying from the stringent requirement set by Applied Biosystems to experiment with a new set of primers capable of amplifying a larger fragment (the 251-bp conservative region of hexon gene). As stated earlier, an amplification length up to 400 bp has been successfully used for real-time PCR applications[48].

A viral seeding study conducted as part of this study indicates that the real-time PCR method is applicable to various environmental matrices. However, the efficiency of amplification is reduced in the presence of high levels of PCR inhibitors. Further improvement of nucleic acid purification could improve the efficiency of quantification and detection.

In conclusion, during the first phase of the study the researchers demonstrated a real-time PCR method for detection of human adenovirus serotype 40. Since the protocol described here does not need any post-PCR steps such as gel electrophoresis, it significantly reduces the problem of PCR carryover contamination. The total time needed to complete a TaqMan PCR run from the time a 96-well plate is put into a thermocycler is 2h 15min. After amplification, the data analyses take approximately 15 min. Thus, as little as 2.5h is needed to complete a TaqMan PCR run and subsequent data analyses.

CHAPTER 3.0

COMPARISON OF REAL-TIME AND NESTED-RCR DETECTION OF ADENOVIRUS IN FECAL CONTAMINATED WATERS

3.1 Introduction

In this chapter, research on comparing the TaqMan real-time quantitative PCR method developed in the previous chapter with the conventional nested PCR is described. The primary goal of the research was to validate the real-time PCR method for applications in environmental samples and to understand the sensitivity and reproducibility of the method.

3.2 Materials and Methods

3.2.1 Sample Preparation

Twelve "blind samples" were prepared by the Southern California Coastal Water Research Project researchers in collaboration with Orange County Sanitation District, Calif., as part of the Microbial Source Tracking (MST) Method Comparison Study project. Each sample was made using sterilized water artificially contaminated with one to three sources of fecal material from either sewage, human feces, dog feces, gull feces, or cattle feces. Sewage or composite single-source fecal material was combined with water to produce a stock solution estimated to contain 10^3 *E. coli* per milliliter of water. Stock solutions were then combined volumetrically to create study samples (Table 3-1). The samples were sent to the lab within 48 hours of preparation.

Table 3-1. Artificially Contaminated Water Samples Used for Study.

Wet Mass or Volume of Feces or Sewage in each 100 ml-Sample					
Sample ID	Sewage	Human	Dog	Cattle	Seagull
	ml	mg	mg	mg	mg
A	0.69				
C		0.89			0.06
E			3.86	96.62	
F		0.22			0.24
G					0.30
I		1.11			
J	0.07		144.93	289.86	
K	0.69				
L		0.67	144.93	48.31	
N	0.14		3.86		
P				386.47	0.06
U				483.09	

3.2.2 Concentration of Viruses

Water samples were concentrated using a Centramate Tangential Flow recirculation ultrafiltration system (Pall Life Science) with a 30 kD molecular weight cutoff Omega filtration cassette following the manufacturer's instruction. The efficiency of viral recovery with this system averaged 55.5%, as determined by a bacteriophage seeding study (Jiang et al., unpublished results). The filtration system was disinfected between each sample with chlorine, followed by flushing with a large volume of DI water. The pH and the membrane cassette flow rate was measured to ensure the absence of chlorine residual and the membrane cassette was regenerated before the system was used for the next sample processing. Sample A, C, E, F, G, I, K, N were concentrated from 10 L to 100 ml. However, Samples J and L were only concentrated from 10 L to 8 L and 4.35 L, respectively, due to the presence of a heavy load of particulates in the samples. Samples P and U were concentrated using Centricon Plus-80 with 100, 000 NMWL membrane (Millipore) from 300 ml to 22 ml. The viral recovery with this system is comparable to the Centramate system with an average phage recovery of 60.4% (Jiang et al. unpublished results). The concentrates were frozen immediately in aliquots. Viral nucleic acid was purified from concentrates using a QIAamp viral nucleic acid purification kit (Qiagen) before use in PCR analysis.

3.2.3 Nested PCR Detection of Human Adenoviruses

Nested PCR for human adenovirus was performed following the protocol of Pina et al [18] with minor modifications. Adenovirus specific primers were 5'-GCCGCAGTGGTCTTACATGCACATC-3' and 5'-CAGCACGCCGCGGATGTCAAAGT-3', yielding an amplicon of 300 bp in size. The nested primers were 5'-GCCACCGAGACGTACTTCAGCCTG-3' and 5'-TTGTACGAGTACGCGGTATCCTCGCGGTC-3', with the resulting amplicon being 143 bp.

For each PCR reaction, 3 µl of sample was used in a 25 µl reaction containing 10 mM Tris-HCl (pH 8.3); 50 mM KCl, 1x TaqM (Eppendorf Inc.), 1.3 mM $MgCl_2$, 200 mM dNTP and 0.1 mM of primers. The PCR was performed using a Hybraid gradient thermal cycler with the following program: 94°C for 4 min, followed by 30 cycles of 92°C for 1.5 min, 55°C for 1.5

min, 72°C for 2 min, and a final extension at 72°C for 10 min. The second round of PCR was performed with the nested primers using the same conditions as above. PCR products were resolved on 2% agarose.

3.2.4 Real-time Quantitative PCR Detection of Human Adenovirus 40

The PCR primers and TaqMan probe designed and tested in Chapter 2.0 were used for this assay. PCR amplification was performed in a 25-µl reaction mixture with a PCR core reagent (Applied Biosystems). The reaction mixture contained 9 µl of viral extract; 1x TaqMan core buffer; 5 mM $MgCl_2$; a 200 µM each of dATP, dCTP, and dGTP; 400 µM dUTP; 300 nM reverse primer and 900 nM forward primer; 250 nM probe; 0.25 U of AmpErase uracil *N*-glycosylase; and 1 U of Taq Gold polymerase. Following activation of the uracil N-glycosylase (2 min, 50°C) and activation of the AmpliTaq Gold for 10 min at 95°C, 40 PCR cycles of 15 s at 95°C and 1 min at 60°C were performed with an ABI 7000 sequence detection system (Applied Biosystems). A five-point standard curve was constructed for each PCR run using serial dilutions of titrated plasmid containing a single Ad40 hexon gene insert. Samples were run in triplicate, and each point of the standard curve was run in quadruplicate. The viral copy number of the unknown sample was calculated from the standard curve. Each run also contained at least three no-template controls to establish the baseline emission intensity of the quenched reporter dye.

3.2.5 Sequence Analysis of Adenoviral Amplicons

Amplicons from nested PCR were cloned into pGEM-T vector (Promega) following manufacturer's instructions. Plasmid DNA with insert was purified and submitted for cycle sequencing using a Dye-Terminator Cycle Sequencing Ready Reaction FS Kit (PE Applied Biosystems) following the manufacturer's instructions. Primers targeted at the T7 and Sp6 promoters were used for bidirectional sequencing. The fragments were trimmed using AssemblyLIGN (Oxford Molecular Ltd.). The final sequences were submitted to a BLAST search of the NCBI GenBank database for identification. ClustalX[49] was used to align sequences. Multiple alignments were performed using default parameters, and the final aligned sequences were refined by eye. A phylogenetic tree was created using the neighbor-joining method[50]. The branching confidence was estimated by bootstrap using the following parameters: random number generator seed—111; number of bootstrap trials—1000. The final tree was viewed in TreeView software.

3.3 Results

The nested PCR using adenovirus-specific primers detected viruses in Samples A, F, K, and N. Sequencing analysis indicated that adenovirus serotype 40 was present in Samples F and N, while the sequences recovered from sample K were identified to be adenovirus sterotype 41 (Figure 3-1). These recovered sequences were over 90% in similarity to the clinical type strain in the GenBank (Table 3-2). Amplicons from Sample A was not successfully sequenced.

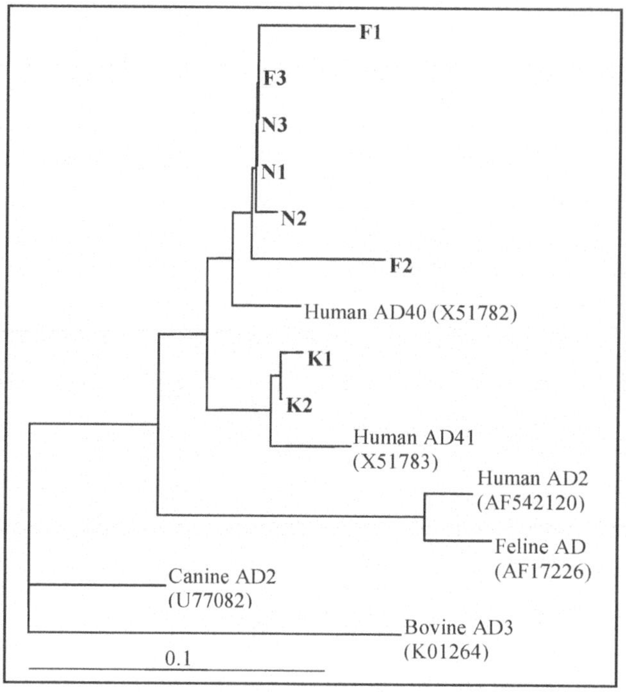

Figure 3-1. Phylogenic Relationships of Viral Sequences Retrieved from Artificially Contaminated Water Samples.

Table 3-2. Similarity of Sequences Retrieved from Artificially Contaminated Water Samples to the Type Strain in GenBank.

Clone Designation	Identification	Similarity
F1	Adenovirus Serotype 40 (X51782)	90%
F2	Adenovirus Serotype 40 (X51782)	97%
F3	Adenovirus Serotype 40 (X51782)	97%
K1	Adenovirus Serotype 41 (X51783)	98%
K2	Adenovirus Serotype 41 (X51783)	97%
N1	Adenovirus Serotype 40 (X51782)	97%
N2	Adenovirus Serotype 40 (X51782)	96%
N3	Adenovirus Serotype 40 (X51782)	97%

The real-time quantitative PCR using primers specific for adneovirus 40 hexon gene quantitatively detected viruses in Samples A and N (Figure 3-2). A total of 40 cycles were run for viral detection in these samples. The amplification curve did not cross the threshold level until Cycle 33 for environmental samples. The standard curve was diluted from 2400 to 2.4 genome copies to covering the range of concentration expected for the environmental samples. Using this standard curve, the viral loads were calculated to be 320 and 120 genome-equivalent per liter of water for Samples A and N, respectively, after correcting for the concentration factor. Only low signal outside the target range of standard curve was found in Samples C, I, J, K, and L, which prevented an accurate quantification of genome copy in the original water samples.

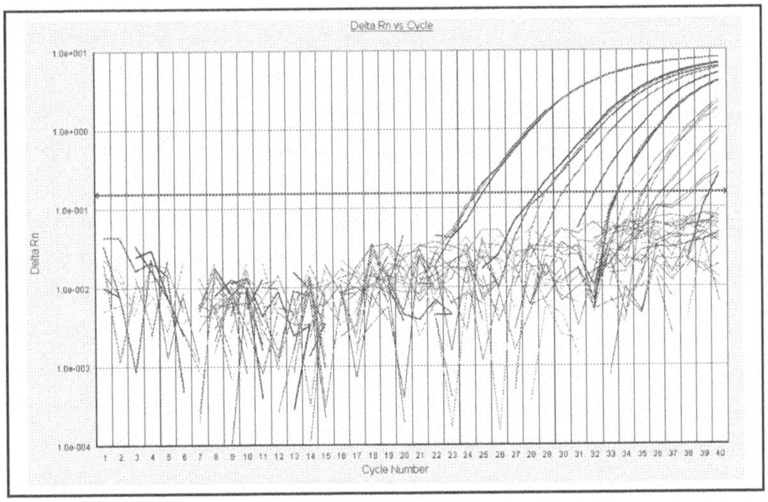

Figure 3-2. Real-time PCR Amplification Plots of Standard Curve and Environmental Samples.

3.4 Discussion

Application of the real-time PCR method to fecal contaminated water samples demonstrated the specificity of the assay. There was no cross-reaction of this assay with fecal sources of nonhuman origin. Real-time PCR yielded positive signals for seven of the eight samples contaminated with sewage and human feces. The negative results may be due to the low concentration of Ad40 in sample 4. Only low signals were detected in samples contaminated with human feces, likely due to the low level of viruses in the individual human feces used for this study. A population of humans should be sampled to yield positive results for human pathogens. Compared with nested PCR assay, which detected adenoviruses in 50% of the samples contaminated with sewage and human feces, this single-step real-time PCR is perhaps more sensitive. However, the quantitative ability of this assay is only limited to the samples with a relatively higher concentration of contaminants. Development of a nested real-time PCR protocol may further improve the sensitivity of the assay. Nevertheless, the real-time PCR method developed here represents a significant advancement in the ability to rapidly quantify viral load in environmental waters.

CHAPTER 4.0

DEVELOPMENT OF REAL-TIME PCR METHOD FOR DETECTION OF ADENOVIRUSES SEROTYPES 2 AND 5

4.1 Introduction

Development and application of real-time PCR primers and TaqMan probe for detection of adneovirus serotype 40 is described in Chapters 2.0 and 3.0. However, a large number of human adenoviruses, especially serotype 2, 5, and 41, which are important human enteric pathogens, are not detectable by this method. Attempts at designing "universal" primers and probe to identify all 51 adenovirus serotypes have not been fruitful. In this chapter the researchers report the investigation of real-time PCR primers and probe for detection of adenovirus serotypes 2 and 5. The primary goal of this research was to identify a set of primer and probe that is complementary to the method described in the previous chapters to monitor a broad spectrum of human adenoviruses in the environment.

4.2 Materials and Methods

4.2.1 Primers and TaqMan Probe

Primers and probe sequences used in the study were those originally described by Pina et al.[18] for the nested step of PCR detection of adenoviruses. The primers are designated as AD5 (F): (GCCGCAGTGGTCTTACATGCACATC) and AD5 (R): (CAGCACGCCGCGGATGTCAAAG). The probe sequence was 6-FAM-CCACCGAGACGTACTTCAGCCTG-TAMRA (located at 101 to123 bp position on adenovirus serotype 5 hexon gene). This primer set, targeting at a portion of the hexon gene, yields an amplicon of 280 bp. The shortest distance between the primer and probe attachment site is 100 bp. Therefore, this set of primer and probe does not fit into the ideal criteria for real-time PCR design as suggested by Applied Biosystems (Forester City, CA).

The specificity of the primers and probe were evaluated against the currently available sequence database using a BLAST program from NCBI (http://www.ncbi.nlm.nih.gov/). The forward primer has homology with human adenovirus serotypes 1, 2, 3, 5, 10, 12, 17, 40, 41, and 48, and the reverse primer is specific for human adenovirus serotypes 2, 4, 5, 7, 8, 9, 15, 16, 17, 19, 22, 37, 40, 41, and 48. Search results also indicate that the probe is specific for four different serotypes of human adenovirus, 1, 2, 5, and 6. There is a three-base-pair mismatch at the 5' end of the probe to adenovirus 41. Therefore, these primers and probe set will specifically detect human adenovirus serotypes 2 and 5 if the PCR condition is set to detect a single base pair mismatch.

4.2.2 Construction of Real-time PCR Standard for Ad5

Using adenovirus 5 specific primers, IPC5F (ATGGCTACCCCTTCGATG) and IPC5R (CTGCTCAGCTTGCTCGTCTA), a 490-bp fragment of Ad5 hexon gene was amplified and subsequently ligated into the pGEM-T Vector (Promega), transferred into *E.coli* XL-1 Blue competent cell, and platted on LB agar containing ampicillin, tetracycline, IPTG (isopropyl-beta-

D-thiogalactopyranoside), and X-Gal (5-bromo-4-chloro-3-indolyl-beta-D-galactopyranoside). White colonies were selected and screened for the presence of insert. The plasmid with insert was designated pAD5 and used as internal positive control (IPC) for optimization of PCR conditions and establishing a standard curve.

4.2.3 Quantification of Ad5 Hexon gene by TaqMan PCR

To test the efficiency of the PCR primers described above and amplification efficiency under various primer concentration and cycle conditions, the primer set was first evaluated using SYBR Green I dye as described in Chapter 2.0.

For TaqMan real-time PCR, amplification was performed in a 25-µl reaction mixture with a TaqMan PCR core reagent (Applied Biosystems). The reaction mixture contained 1 ng of plasmid pAD5 purified by the QIAgen miniprep kit (Qiagen); 2x TaqMan core buffer; 5 mM $MgCl_2$; a 200-µM concentration (each) of dATP, dCTP, and dGTP; 400 µM dUTP; a 900 nM concentration of each primer; 300 nM probe (see below for primers and probe optimization); 0.25 U of AmpErase uracil N-glycosylase; and 1 U of Taq Gold polymerase. Following activation of the uracil N-glycosylase (2 min, 50°C) and activation of the AmpliTaq Gold for 10 min at 95°C, 40 cycles consisting of 15 s at 95°C and 1 min at 55°C were performed with an ABI 7000 sequence detector system (Applied Biosystems). Each run contained at least three no-template controls to establish the baseline emission intensity of the quenched reporter dye.

4.2.4 Construction of Standard Curve

Standard curves were generated by using serial dilutions of known amounts of nonlinearized plasmid pAD5. All samples were run in quadruplicates, and positive and negative controls were included at each run. The amount of DNA was defined as the median of the quadruplicate data obtained.

4.2.5 Optimizing Primers and Probe Concentration for Real-time PCR

The procedure for optimizing primer and probe concentration was essentially the same as described in Chapter 2.0. The concentration matrix for primers and probe was established by varying the concentration of primers or probe at each time.

4.3 Results

4.3.1 Specificity of Primers and Optimization of PCR Conditions

Amplification efficiency and specificity of the primers were first tested using SYBR Green dye as described in Chapter 2.0 for adenovirus serotype 40. Comparable results were achieved using this new primer set for adenovirus serotype 5 (Ad5). Steady increases of fluorescence signals in samples containing various concentrations of Ad5 virus were observed in comparison with background and no-template control. This increase of fluorescence signal was also proportional to the initial concentration of the target, suggesting amplification of the desired target. A dissociation curve analysis was run following each SYBR Green PCR assay showing a single peak at the expected Tm temperature (Figure 4-1). A single dissociation peak with a Tm within a variation of 2°C of the expected temperature was observed, suggesting amplification of a specific product[38].

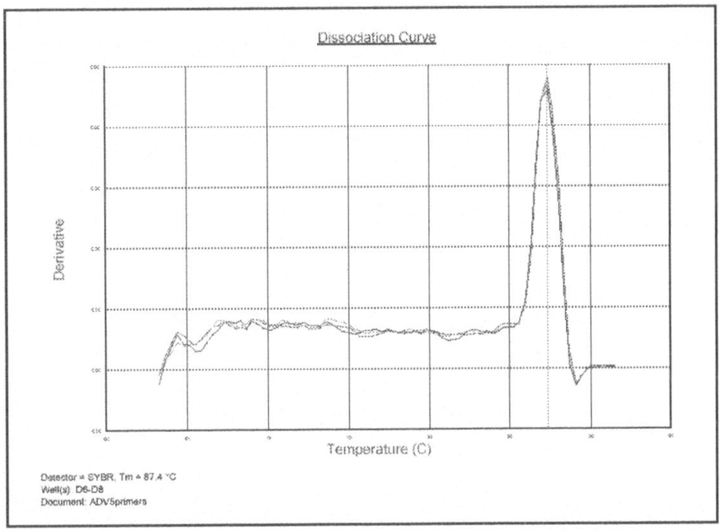

Figure 4-1. Dissociation Analysis of Adenovirus Serotype 5 Real-time PCR with SYBR Green.

The results of primers and probe concentration optimization are shown in Figure 4-2. A probe and primer concentration combination of 300 nM forward and 900 nM reverse primer yielded the equivalent Ct value to the highest concentration used (900/900 nM); therefore this combination was used for all future assays.

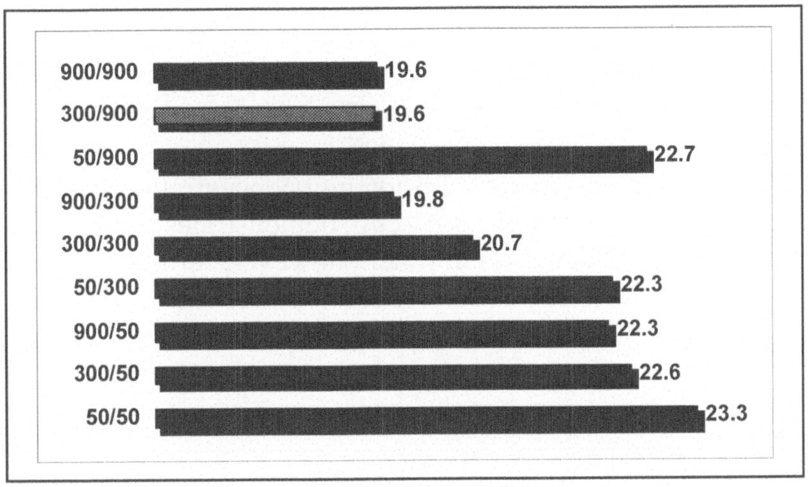

Figure 4-2. Primers and Probe Concentration Optimization.
Numbers on the left indicate concentrations (nM) of primers used in combination. Bar length Indicates threshold value.

4.3.2 Standard Curve

Using serial dilutions of pAd5, the standard curve generated from the PCR data resulted in a reaction efficiency of >90%, with correlation coefficients of 0.990436 and a slope of -4.33 (Figure 4-3). The standard curve is linear over a 5-log concentration, suggesting the PCR primers and probe are suitable for accurate quantification of viral genome loads.

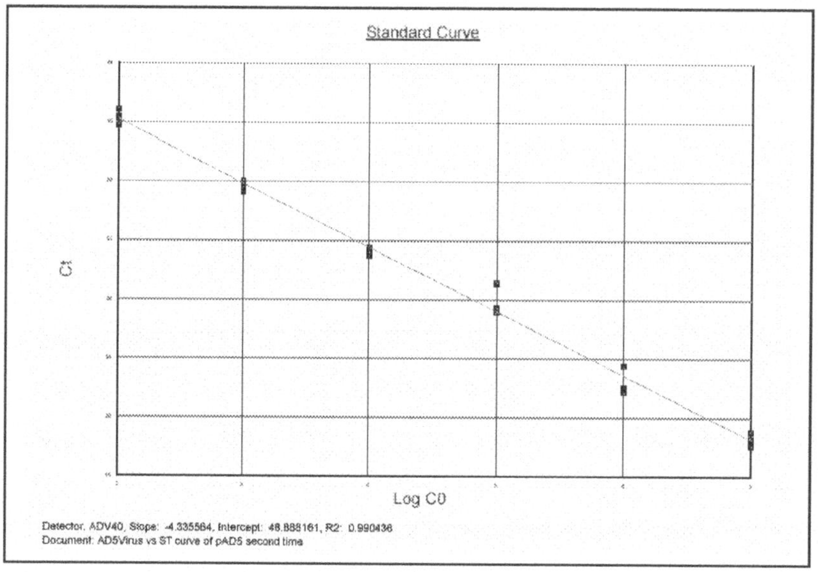

Figurer 4-3. Real-time Quantification of pAd5 Stand Curve.

4.4 Discussion and Conclusions

Although the design of primers and TaqMan probe was not ideal according to manufacturer's recommendation for real-time PCR application, the results of this study showed that primers and probe presented here are suitable for detection and quantification of Ad5 viral genomes in samples that contain adenoviruses over a 6-log unit concentration range. The method tested in this study, together with those developed for Ad40, demonstrates the potential utility of real-time PCR for quantification of viral genomes in water. However, further development is necessary to apply a single assay for detection of diverse serotypes of adenovirus serotypes.

CHAPTER 5.0

REAL-TIME PCR DETECTION OF HEPATITIS A VIRUS

5.1 Introduction

This chapter discusses the progress in the development of a real-time RT-PCR method for detection of human hepatitis A virus (HAV). HAV is a hepatotropic member of the *Picornaviridae* family. The clinical manifestations of HAV infection in humans can vary greatly, ranging from asymptomatic infection, commonly seen in young children, to fulminant hepatitis, which in some cases can result in death. HAV is transmitted primarily by the fecal-oral route, and epidemics are common in regions where sanitation is poor. Hepatitis A continues to be one of the most frequently reported vaccine-preventable diseases in the U.S.

Although hepatitis A occurs in virtually every area of the U.S., western states including Southern California have higher rates than rest of the country http://www.cdc.gov/ncidod/diseases/hepatitis/a/vax/index.htm). The viron is nonenveloped with a capsid composed of three main polypeptides (VP1, VP2, and VP3). Only one serotype of HAV has been identified thus far, and the only known antigenic variants are HAV strains collected from Old World monkeys[51-56]. A previous study reported at American Society for Microbiology annual conference presented a method for real-time RT-PCR detection for HAV[57]. The method reported here, although similar to the previous study, targets a different region of the viral genome. This method, although preliminary, may provide an alternative approach to end-users for rapid detection of HAV in environmental samples.

5.2 Materials and Methods

5.2.1 Virus Stocks

The HAV HM175 used in this study was provided by Dr. Marylynn Yates, Department of Environmental Sciences, University of California, Riverside. The concentration of virons was not determined.

5.2.2 Primers and Probe

Primers and probe were designed using Primer Express 2.0 software (Applied Biosystems) and were directed to the highly conserved VP3 capsid gene of the HAV genome (Table 5-1). The reporter (FAM) and quencher dyes (TAMRA) were attached to the 5' end and 3' end of the probe, respectively. The criteria for primer and probe design were essentially the same as those described in Chapter 1.0. The specificity of the primers was first tested using SYBR Green real-time PCR as described in Chapter 2.0 before experiments were carried forward to TaqMan PCR for absolute quantification.

Table 5-1. Primers and Probe Sequences Used in the HAV Real-time TaqMan PCR.

Target	Primers	Probes	Accession
HAV	HEPAF: TGGTGGGATCAAAATTACTC HEPAR: AATTTGTTGACCAACTGAGG Product size: 100 bp (1598-1697)	6-FAM-TCCAACTTTGGCTGCTCAGTTTC-TAMRA (1640-1662)	HAV; HM175 M14707
IPC[1] HAV	PHAV1: CAGCACATCAGAAAGGTGAG PHAV2: CTCCAGAATCATCTCCAAC Product size: 192 bp (2035-2226)	N/A	HAV; HM175 M14707

[1*]Internal Positive Control
6-FAM; 6-carboxyfluorescein
TAMRA; 6-carboxytetramethyl rhodamine.

5.2.3 Construction of the Real-time PCR Internal Standard

The viral RNA was extracted from a viral stock HAV HM175 using the QIA viral RNA purification kit (Qiagen). The RNA then was reverse transcribed using MultiScribe reverse transcriptase and random hexamers (GeneAmp RNA PCR, Applied Biosystems) to cDNA. For reverse transcription, each 10-µl reaction contained 3 µl of RNA, 1 µl of 10x RT buffer, 5 mM $MgCl_2$, a 500-µM concentration of each of the deoxynucleoside triphosphates, a 2.5-µM concentration of random hexamer, 2.5 U/µl of MultiScribe RT, and 1 U/µl of RNase inhibitor. cDNA synthesis was performed in an automated thermocycler PCR-Express (Hybaid, Ashford, United Kingdom) using the following protocol: initial step of 99°C for 5 min, followed by an annealing step at 25°C for 10 min; reverse transcription was carried out at 42°C for 30 min, followed by reverse transcriptase inactivation at 95°C for 5 min. Using SC HAV primers (Table 5-1) and cDNA, a 192 bp of highly conserved region of the VP3 capsid gene of the HAV was amplified and subsequently ligated into the pGEM-T Vector (Promega), transferred into *E.coli* XL-1 Blue competent cell, and platted on LB agar containing ampicillin, tetracycline, IPTG (isopropyl-beta-D-thiogalactopyranoside), and X-Gal (5-bromo-4-chloro-3-indolyl-beta-D-galactopyranoside) as described in Chapter 2.0. White colonies were selected and analyzed for the presence of insert. In order to determine the orientation of insert, selective clones of the recombinant plasmid (designated pHAV) were sequenced using a Dye-Terminator Cycle Sequencing Ready Reaction FS Kit (PE Applied Biosystems, Foster City, Calif.) following the manufacturer's instructions. Primers targeted at the T7 (TAATACGACTCACTATAGGG) and Sp6 (ATTTAGGTGACACTATAGAA) promoters were used for bidirectional sequencing. Sequencing analysis revealed that the insert is cloned in front of the SP6 promoter. Using the sequences in the web cutter site (http://www.firstmarket.com/cutter/cut2.html), the restriction map of the insert and the vector were determined, and the enzyme that only introduced one cut (*Not*I) was selected for in vitro transcription.

5.2.4 In Vitro Transcription

pHAV was linearized with *Not*I and gel purified. The purified linear pHAV was used as a template for RNA transcription using the promoter for SP6 RNA polymerase (RiboMAX™ Large Scale RNA Production Systems, Promega). Each 100 µl of reaction contained SP6 5 x transcription buffer (20 µl), 20 µl rNTPs (25mM ATP, CTP, GTP, UTP), 50 µl linear DNA template (5–10µg total), and 10 µl SP6 enzyme mix. The reaction mix was incubated at 37°C for 2–4 hours. Any contaminating DNA was eliminated using the RNase-free DNase I enzyme supplied with the kit, which was then destroyed by heating to 94°C for 5min. After removal of the DNA template and unincorporated nucleotides according to the kit supplier's instruction (Promega), 2 µl of RNA, 20 µl of RNA sample buffer (supplied with the kit), and 5 µl of RNA loading buffer (50% glycerol, 1 mM EDTA, 0.4% bromophenol blue, 1 mg/ml ethidium bromide) were mixed and heated for 10 min at 65°-70°C. The RNA transcript size was determined in 2% agarose gel in 1x TAE containing 0.5µg ethidium bromide.

5.2.5 Real-time RT-PCR Assay

Assay conditions were optimized in an ABI Prism 7000 sequence detection system (Applied Biosystems); 10 µl of RNA extract was added to the PCR reaction mixture containing a final concentration of 300 nM forward primer, 900 nM reverse primer, 150 nM dual-labeled fluorogenic probe, 1× TaqMan universal master mix, and 1× RT mix (One-Step RT-PCR TaqMan reagent kit from Applied Biosystems). RNA extracts were substituted by nuclease-free water in negative control reactions. Samples were assayed in duplicate reactions in a total volume of 25µl. RT was performed at 48°C for 30 min and immediately followed by activation of AmpliTaq Gold DNA polymerase at 95°C for 10 min. A total of 40 cycles was performed, consisting of a denaturation step at 95°C for 15 s and a combined annealing-extension step at 60°C for 1 min. Fluorescence data were collected during each annealing-extension step and analyzed using ABI Prism SDS software Version 2.0.

5.3 Results and Discussion

During this study, the researchers constructed an internal RNA standard using reverse transcription of cloned viral cDNA. This step is critical for reproducible results of quantitative real-time RT-PCR assay. Optimal RNA yields from reverse transcription depend on the purity and quantity of DNA template. No RNase should be present in the DNA preparation, and templates must be linear prior to in vitro transcription to produce RNA of defined length. It is also important to avoid using restriction enzymes that produce 3′ overhangs. Extraneous transcripts have been reported to appear, in addition to the expected transcript, when such templates are transcribed. The extraneous transcripts can contain sequences complementary to the expected transcript, as well as sequences corresponding to the vector DNA.

The results of SYBR Green real-time RT-PCR using this RNA standard is shown in Figure 5-1. A single peak of dissociation was found from the PCR product, suggesting that the primers used for this assay are specific for HAV detection. The efficiency of TaqMan real-time PCR using purified RNA standards is comparable to previous study for Ad40 and Ad5. However, RNA standards were very unstable in comparison with adenovirus DNA standards. The quantity of RNA transcripts degrades even under frozen storage condition. Thus, only freshly prepared RNA transcripts can be used for accurate quantification of unknown samples, which makes calibration of a standard curve for each real-time PCR reaction tedious. Alternative calibration standards for real-time quantification should be explored to achieve stable results for environmental application. Since several other research labs within the U.S. are also currently undertaking the study for real-time quantification of HAV, Project Subcommittee members suggested devoting the current research to adenoviruses rather than duplicating the effort for HAV quantification. The research on real-time PCR quantification of HAV was not carried any further. The project team hopes the preliminary data and pitfalls reported here can be used as the starting point for other researchers to further investigate HAV quantification.

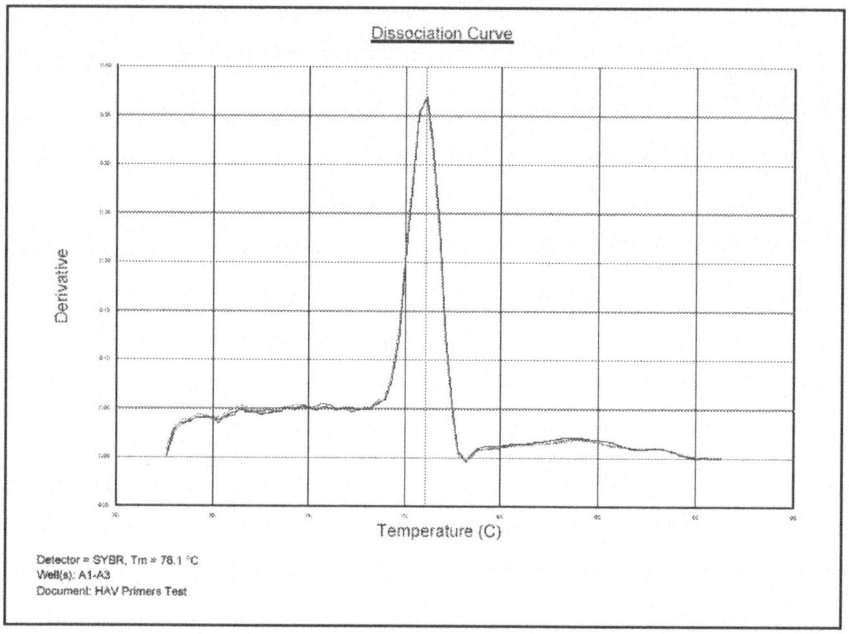

Figure 5-1. Dissociation Curve of Real-time RT-PCR of HAV Transcripts with SYBR Green I.

CHAPTER 6.0

SEQUENCE ANALYSIS OF ENVIRONMENTAL ADENOVIRUSES

6.1 Introduction

This chapter addresses the concerns and recommendations of Project Subcommittee members from a previous review about the isolating and sequencing of environmental adenoviruses. Since the overall goal of the project was to identify an index organism for human viral contamination, it is important to understand the serotypes and genetic diversity of adenoviruses in the environment. The review of literatures in clinical research indicates that nearly 50% of adenoviruses shed in human feces belong to serotypes 40 and 41, while adenovirus serotype 5 is more prevalent in adult populations because it causes persistent asymptomatic infection of adenoid gland[58]. This chapter reports on the effort to isolate environmental adenoviruses using tissue culture and subsequent amplification, cloning, and sequencing of adenovirus hexon gene for serotype identification.

6.2 Materials and Methods

6.2.1 Culture of Environmental Adenoviruses

Water samples were collected from 15 locations in Newport Bay, Calif., a small estuary enclosed by a highly urbanized area. Five hundred milliliters of water from each site were concentrated using a Centri-Plus 80 ultrafiltration concentration system with a 100 KD molecular-weight-cutoff membrane as described in the Chapter 3.0. The concentrates were purified by chloroform extraction and filtered through 0.2-µm pore-size low-protein-binding syringe filters. The filtrates were used for isolation of human adenoviruses.

Two tissue culture cells lines were used for isolation of infectious human viruses. The human embryonic kidney (HEK- 293A) cell line was obtained from University of Southern California (courtesy of Professor Michael Lai) at 34-passage and was used in this study between 42 and 50. The cells were grown at 37°C in Dulbecco's Modified Essential Medium (DMEM), containing 5% fetal calf serum (FBS), 2 mM L-glutamine, 100 U/ml penicillin and 100 mcg/ml streptomycin. The human lung carcinoma cell line (A549) was obtained from the Los Angeles Sanitation District at 108-passage and was used in this study between 115 to 123. A549 cells were grown in Ham's F12 medium, also containing 5% FBS, 2 mM L-glutamine, 100 U/ml penicillin, and 100 mcg/ml streptomycin. Cells were subcultured at four- to five-day intervals with a trypsin EDTA solution (Cellgro). Both HEK-293A and A549 cell lines are known to support the growth of human adenoviruses.

Plaque assays were performed using concentrated and purified Newport Bay samples after extraction with 1 volume of chloroform to remove cytotoxicants, clarified by filtering through 0.2-µm pore size filter (PES, 4mm, low protein binding, Whatman). Samples were equally inoculated onto confluent HEK-293A and A549 cells in 6-well plates. The plates were incubated in a CO_2 incubator at 37°C for 1 hour and were rocked every 15 min. Inoculated cells were then washed once with pre-warm calcium and magnesium-free PBS (pH 7.4, Mediatech, Inc.) and overlaid with DMEM 5% FBS and 1.25% agarose containing 50 mg/ml gentamicin and 0.25 mg/ml amphotericin B. Cells were incubated at 37°C, 5% CO_2. A second overlay with the same contents as above was applied on the sixth or seventh day. Inoculated cell cultures were examined daily under a light microscope for up to two weeks post-infection. The cells with viral infection were harvested by gently scraping off the agarose layer and collecting the cells in PBS buffer.

6.2.2 Amplification, Cloning and Sequencing of Environmental Adenoviruses

Adenoviruses were released from infected cells using multiple rounds of freeze-thaw cycle. Viral DNA was extracted using QIAViral Purification Kit (Qiagen). The adenovirus hexon gene was amplified using a degenerate primer set Ad-1 and Ad-2 as previously described by Xu et al. (2000). The primer pair matches sequences for eight adenovirus serotypes although it is possible to amplify additional serogroups as claimed by the original authors[59]. The resulting amplicon is 482 bp. The researchers chose this primer set instead of the primers used in Chapter 3.0 for nested PCR because this primer set is capable of capturing a broad spectrum of environmental adenoviruses. This set of primer also yields a 600 bp amplicon for diversity analysis and phylogeny comparison.

Amplicons were purified from the gel using QIA gel purification kit (Qiagen) and cloned into *E. coli* using TOPO TA cloning kit (Invitrogen Inc.) following manufacturer's recommendation. The clones with insert were screened using restriction analysis of plasmid prep. Cycle sequencing was performed using primer target at T7 promoter sequence as described in Chapter 3.0.

6.2.3 Sequence Analysis

Resulting sequences were trimmed and submitted for a BLAST search on the NCBI database. The resulting sequence that displays a greater than 99% gene sequence similarity to an existing gene sequence in the database was considered the same serotype.

6.3 Results

In the Newport Bay samples, viral plaques were only observed on the HEK-293A cell line. Plaques were developed between four to eight days with diverse plaque morphologies. Nucleic acid extraction from tissue culture lysates were only amplifiable using primers specific for adenoviruses but not with primers specific for enteroviruses, suggesting most of the plaques observed on the HEK-293A cell lines were adenoviruses. Forty-four randomly selected clones of 482 bp adenovirus PCR amplicon were sequenced. Sequence alignment with the current GenBank (NCBI) database identified that 55% of the environmental isolates belonged to serotype 40 while the other 45% of the clones were more similar to serotype 5 (Figure 6-1a,b). The similarity with clinical strains ranged from 93% to over 99%, suggesting that these environmental isolates were of human origin.

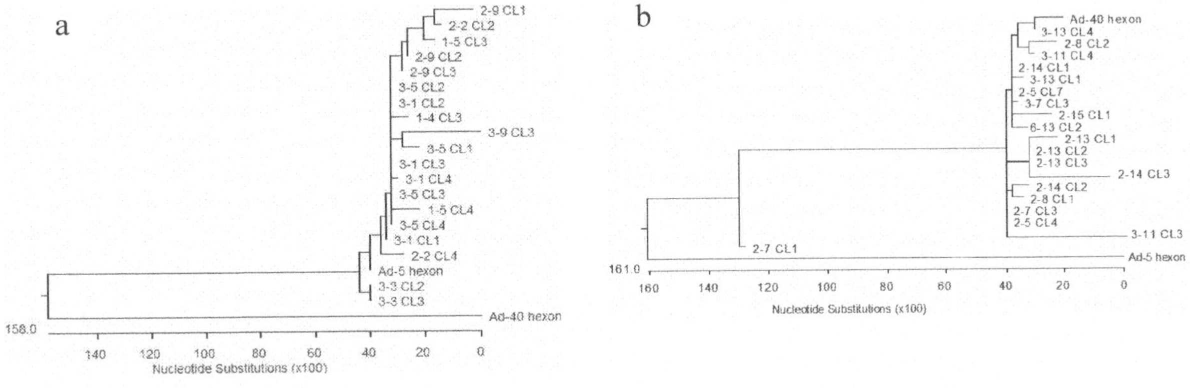

Figure 6-1. Genotype of Adenoviruses Recovered from Newport Bay, Calif. by Plaque Assay on HEK-293A Cell Line.

6.4 Discussion and Conclusions

The sequences recovered from environmental isolates are similar to those in clinical collection in the database, suggesting that environmental adenovirus isolates are of human origin. This result confirmed the clinical observation that adenovirus serotypes 40 and 41 are the major serotypes shed in human feces. However, the sequences recovered are more similar to serotype 40 than 41. There is only 6-bp mismatch between serotype 40 and 41 within the amplicon region. These differences may not sufficiently differentiate serotype 40 from 41. The sequence obtained from this study, however, covers only a very small fraction of the adenovirus genome. Furthermore, the hexon gene used in this study may not be a good representation of the phylogenic or evolutionary relationship between adenoviruses isolated from the environment. The results of sequence analysis presented here should be considered solely as a confirmation of the similarity between environmental adenoviruses and their clinical counterparts.

It is also not surprising to find an important portion of adenovirus serotype 5, as this serogroup causes a broad spectrum of diseases in children and asymptomatic infections in adults. However, viral plaques were recovered from 293A cell line but not A549 cell line. A549 was known to support the proliferation of adenovirus serotype 40, but the adenovirus isolated from Newport Bay may be different from the serotype 40 used on A549 cell line. Further investigation of host specificity will contribute to the understanding of adenoviruses in the environment and their infectivity. In summary, the results suggest that the area sampled at the Newport Bay estuary may be contaminated with human waste and that human serotypes 40, 41, and 5 are important serotypes that may be found in the polluted environment. The results from this study will be used as a guideline for future development of adenovirus real-time PCR assay for environmental samples.

CHAPTER 7.0

DEVELOPMENT OF DEGENERATE PRIMERS AND PROBE FOR REAL-TIME PCR DETECTION OF A BROAD SPECTRUM OF ADENOVIRUS SEROTYPES

7.1 Introduction

Previous chapters reported the development of a real-time PCR method for detection of adenovirus serotypes 40, 2, and 5. However, sequence analysis of environmental adenovirus as described in Chapter 6.0 suggests that serotypes 40, 41, and 5 are prevalent in the environment. In this chapter, design and testing of degenerate primers for real-time PCR detection of a broad spectrum of adenovirus serotypes, including 5, 40, and 41 in a single reaction are described.

7.2 Materials and Methods

7.2.1 Primers and Probe

The primers and probe used in this study are listed in the Table 7-1. Primer pair Ad-1 and Ad-2 were those originally designed by Xu et al.[59]. Ad-3, Ad-4, and degenerate probe Adprobe7 were designed in this study using multiple sequence alignment. Figure 7-2 shows a schematic graph of the location of primers and probe on the adenovirus hexon gene. The combination of primer Ad-1/Ad-2 yields a predicted product of 482 bp and specific for eight different steroptypes; Ad-3/Ad-2 yields an amplicon of 163 bp with specificity to seven different serotypes; Ad-4/Ad-2 gives an amplicon of 116 bp with specificity to eight different serotypes. The probe sequence is located close to Ad-2 and covers all serotypes that are amplifiable by primer pairs and more.

Table 7-1. Primer and Probe Sequences Used for Real-time PCR Detection.

Primer Pairs	Sequence	Targeted Adenovirus Serotypes
Ad-1/Ad-2	Ad-1: TTC CCC ATG GCI CAY AAC AC	2, 3, 5, 7, 17, 21, 40, and 41
	Ad-2: CCC TGG TAK CCR ATR TTG TA	
Ad-3/Ad-2	Ad-3: GAC TCY TCW GTS AGY TGG CC	3, 5, 7, 17, 21, 40, and 41
Ad-4/Ad-2	Ad-4: GAG TTY GAA ATC AAR CGC AC	2, 3, 5, 7, 17, 21, 40, and 41
Adprobe7	6-FAM- AGR AAC CAG TCY TTG GTC ATG TTR CAT TG-TRMRA	1-5, 7-9, 17, 19, 21, 28, 34, 37, 40, 41, 47, and 48

Adenovirus Hexon Gene (2.9 kb)

Amplicon
Ad-1/2: 482 bp
Ad-3/2: 163 bp
Ad-4/2: 116 bp

Figure 7-1. Primers and Probe Location on Hexon Gene.

7.2.2 Quantification Standards for Real-time PCR

To create adenovirus real-time PCR standards, Ad40 hexon gene was amplified by a pair of degenerate primers, Ad1: 5'-TTC CCC ATG GCI AYA ACA C-3' and Ad2: 5'-CCC TGG TAK CCR ATR TTG TA-3' (Xu et al. 2000). The resultant 482-bp amplicon was cloned into the pCR2.1-TOPO vector (Invitrogen Inc.) and confirmed by sequencing. The purified recombinant plamid DNA was quantified by UV spectrophotometer and then serially diluted in double distilled H_2O to a final concentration ranging from 2.5 x 10^3 to 2.5 x 10^8 copies of genome equivalent/ml. Four-microliter aliquots of each dilution (10 to 10^6 genome equivalent /reaction) were used for real-time PCR in triplicates to create the standard curve and used as quantification standards for adenovirus genome copy in experimental samples. A new standard curve was run for each real-time PCR reaction.

7.2.3 Optimization of Primer Combinations

The goal of these experiments was to identify the most efficient primer pairs for target amplification. To fulfill the needs of small amplicon size for efficient TaqMan real-time PCR, the researchers chose two new primers, Ad-3 and Ad-4, in combination with the existing primer Ad-2[59]. The PCR efficiency was first tested using SYBR Green I method. All assays were performed using clones of adenovirus 40 hexon gene as targets. PCR assays were performed using the manufacturer's default setting: 95°C 15 sec and 60°C 1 min for 40 cycles. The PCR reaction was performed using the ABI real-time PCR core buffer system as described in detail in Chapter 2.0.

7.2.4 Optimization of TaqMan PCR Conditions

To increase the efficient of amplification with TaqMan probe, the researchers modified the manufacturer's default setting for TaqMan PCR by adding an annealing step between 55° to 56°C and adjusted the extension temperature to 62°C. The following cycling conditions were tested in addition to the manufacturer's default: 1) 95°C 15", 62°C 1' for 40 cycles; 2) 95°C 15", 55°C 15", 62°C 1' for 40 cycles; 3) 95°C 15", 56°C 15", 62°C 1' for 40 cycles; 4) 95°C 15", 56.5°C 15", 62°C 1" for 40 cycles; 5) 95°C 15", 56°C 15", 62°C 40" with an increase of 1" per cycle, for a total of 45 cycles. They also tested the efficiency of different primer combinations with Adprobe7 for amplification of the target. Primer and probe concentrations were optimized using the matrix described in Chapter 2.0. The sensitivity of the real-time PCR was tested using a dilution of the linear target DNA as described in detail in Chapter 2.0.

7.2.5 Real-time PCR Detection of Adenoviruses in Sewage Using Degenerated Primers

Sewage samples were collected by random sampling in two Southern California sewage treatment plants. Primary effluent was raw domestic sewage after sedimentation. Secondary effluent was treated by activated sludge. Some of the secondary effluents were also treated with chlorine at various locations in the effluent transportation pipeline. The final concentration of chlorine residue was between 2-3 ppm in the composite samples.

For real-time PCR of adenoviruses, 10-ml sewage samples were ultracentrifuged at 207,750 x g for 90 min. Pellets were resuspended in 500 µl supernatant and extracted using equal volume of chloroform twice. Viral nucleic acid was extracted from the concentrates using QIAamp viral RNA Mini Kit (Qiagen) following the manufacturer's instruction and finally eluted in 30 ml ddH2O. Real-time PCR was carried out in 25-µl reaction mixtures consisting of 1x TaqMan master mix, primers, TaqMan probe, and template DNA using the optimal condition developed above. The final thermocycling profile for adenoviruses was 95°C 15 sec, 56°C 15 sec, and 62°C 30 sec with 1 sec auto increment every cycle for 45 cycles. All samples were run in triplicate.

7.2.6 Adenovirus Plaque Assay in Sewage

Ten-milliliter sewage samples were ultracentrifugated at 207,570 x g for 90 min. Pellets were resuspended in 500 µl phosphate buffered saline (PBS, pH7.4) and extracted with one volume of chloroform twice to remove cytotoxicants. The supernatants were further clarified by filtering through 0.2 µm-pore size filters (PES, 4mm, low protein binding, Whatman). An equal portion of each purified sample was inoculated onto confluent HEK-293A and A549 cells in 6-well plates. The plates were incubated in a CO_2 incubator at 37°C for 1 hour and were rocked every 15 min for viral adsorption. Then, inoculated cells were carefully washed once with pre-warmed PBS (pH 7.4) and overlaid with warm DMEM 5% FBS and 1.25% agarose containing 50 mg/ml gentamicin and 0.25 mg/ml amphotericin B. Cells were incubated at 37°C with 5% CO_2. A second overlay was applied at six to seven days post-infection (p.i). Inoculated cell cultures were examined microscopically daily up to 2 weeks p.i. Plaques were counted as plaque forming unit (PFU)/100 ml sample at 10 days. Highly diluted recombinant adenovirus serotype 5 was used on HEK-293 cells as a positive control, with PBS as a negative control. The recombinant adenovirus expresses green firefly protein in infected cells, so the resultant plaque can be confirmed and differentiated under UV-microscope.

7.3 Results and Discussion

7.3.1 Selection of Primers and PCR Conditions

The effect of different primer combinations on the results of PCR amplification is shown in Figure 7-2. All primer combinations are capable of producing the expected amplicons. However, primer pair Ad-1/Ad-2 appears most efficient when SYBR Green I was used in the absence of TaqMan probe, while the Ad-4/Ad-2 combination was low in amplification efficiency at the default PCR cycle setting. This result is most likely due to the large size of the amplicon produced by Ad-1/Ad-2 and the presence of nonspecific amplification products that are detected by SYBR Green dye as the target product (Figure 7-3).

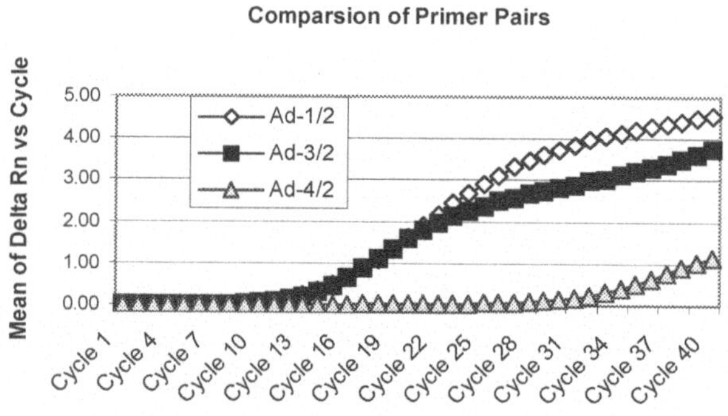

Figure 7-2. Comparison of Primer Pairs for Amplification Efficiency.

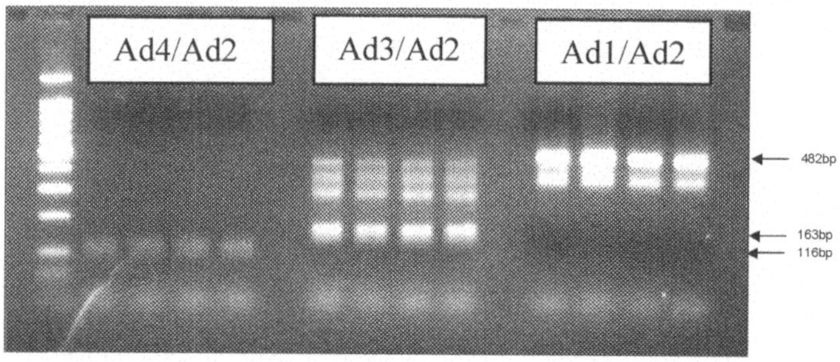

Figure 7-3. Amplification Products of Different Primer Pairs.

To increase the amplification efficiency and specificity, an additional annealing step at 55°C was added in the presence of the TaqMan probe (Figure 7-4). This step significantly reduced the nonspecific products from the reaction. The results indicated that primer pair Ad-3/Ad-2 in combination with TaqMan probe Adprobe 7 were the most robust. Because the TaqMan probe is only specific to the target template, this reaction theoretically eliminates the interferences from nonspecific amplification for the final product quantification.

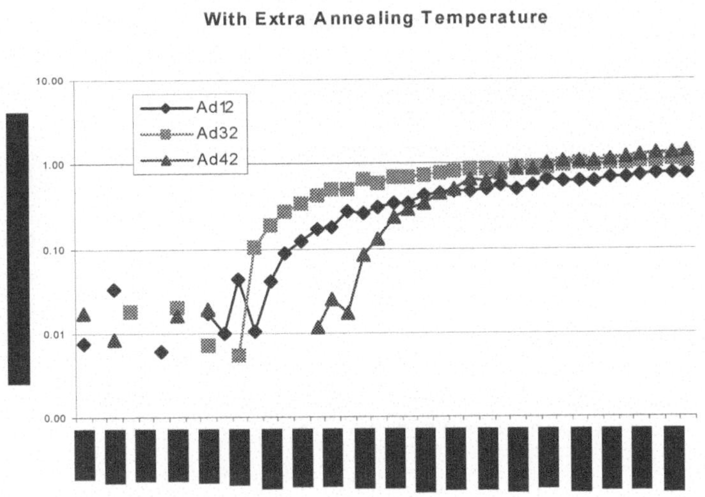

Figure 7-4. Optimization of TaqMan Real-time PCR with Different Primer Combinations and an Additional Annealing Step at 55°C (95°C 15", 55°C 15", 62°C 1" for 40 Cycles).

7.3.2 Optimization of Primers and Probe Concentrations

Figures 7-5 and 7-6 show the results of primers and probe concentration optimization. The best amplification efficiency was found at 300 nM/300 nM for primers, and the probe concentration of 200 nM was the most efficient. Therefore, these conditions were used for constructing standard curves. Figure 7-7 shows a standard curve with R^2 value of 0.95 and slope of –5.79.

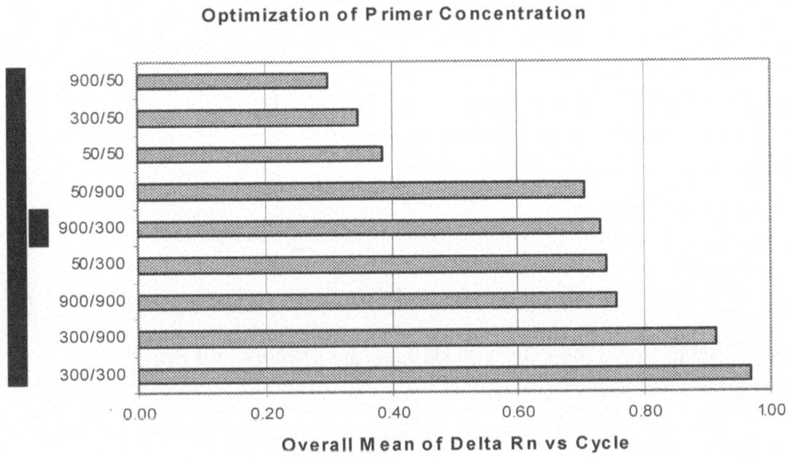

Figure 7-5. Optimization of TaqMan PCR Primer Concentrations.

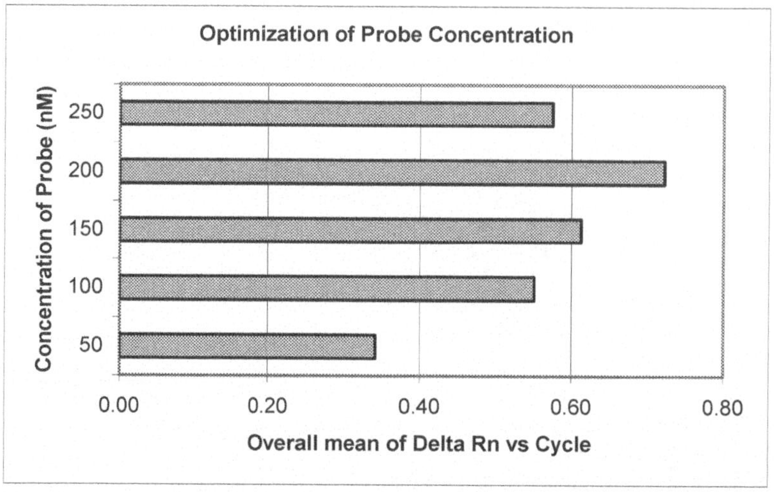

Figure 7-6. Optimization of TaqMan Probe Concentration.

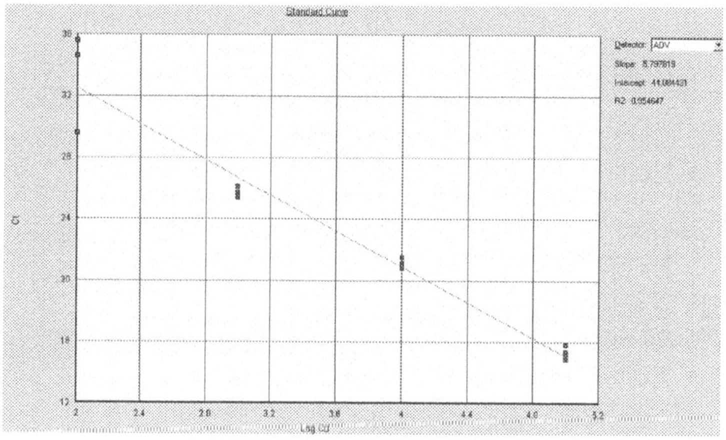

Figure 7-7. Standard Curve for TaqMan Real-time PCR with Degenerate Primers and Probe.

7.3.3 Optimization of Annealing Temperature

Figure 7-8 shows the results of optimization of annealing temperature. Annealing temperature at 56°C appears to be more efficient for the target amplification than 55°C.

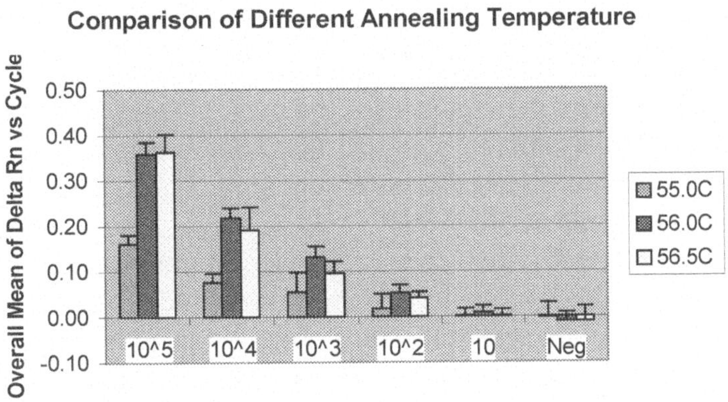

Figure 7-8. Optimization of Annealing Temperature.

7.3.4 Application of Real-time PCR for Quantification of Adenoviruses in Sewage

Real-time PCR with the degenerate primers and probe designed above was applied to quantify adenoviruses in sewage samples and the results were compared with those from plaque assays on A549 and HEK-293A cell lines. Typical viral plaques with diverse morphology were observed on HEK-293A cells but not on A549 cells, indicating the HEK-293A cell line may be more sensitive to infection of adenovirus than A549. Interestingly, the number of infectious adenoviruses decreased from 67 or 80 PFU/100 ml in the primary sewage effluents to below detection in the secondary effluents in samples collected from both treatment plants (Table 7-2). However, there was virtually no difference in viral load between primary effluents and secondary effluents detected by real-time PCR (Table 7-2). The average number of adenoviral genomes detected in sewage by real-time PCR was 8.1×10^4 viral genomic copies/100 ml, which is comparable to the genomic copies of poliovirus in sewage reported by Tsai et al.[60].

Table 7-2. Comparison of Adenovirus Quantification by Real-time PCR and Plaque Assay in Sewage Samples.

Sewage	Real-time PCR (Viral genomic copies/100 ml)	Plaque Assay (PFU/100 ml)
9/30/03		
Primary effluent–Plant I	$6.6x10^4$	67
Secondary effluent–Plant I	$6.1x10^4$	Not detected
Primary effluent–Plant II	$7.0x10^4$	Not detected
Secondary effluent with chlorine–Plant II	$8.5x10^4$	Not detected
10/1/03		
Primary effluent–Plant I	$7.4x10^4$	80
Secondary effluent–Plant I	$7.4x10^4$	Not detected
Secondary effluent with chlorine–Plant II	$14x10^4$	Not detected

7.4 Conclusions

The results suggest that the degenerate primers and probe designed in this study are capable of real-time quantification of adenoviruses in environmental samples. The viral quantity determined by real-time PCR is significantly higher than that detected by plaque assay. This result is not a surprise because real-time PCR detects genome equivalence of adenovirus. It is well known that there is a large range of variability between viral particles and plaque-forming unit. Genome equivalence detected by real-time PCR may be more similar to particle numbers than plaque-forming units. In addition, the environmental degradation condition may further reduce the viral infectivity without completely destroying the genome targeted by real-time PCR assay. This is evident through the consistent detection of a similar genome number in primary as well as secondary treated sewage, while infectivity of adenoviruses in the secondary treated sewage was lower by plaque assay. Overall, the result of this study raised the question of how to interpret the molecular biology-based viral quantification results and their implications for human health risks. In the next step, the researchers applied this method to the field samples for adenovirus detection.

CHAPTER 8.0

IMMUNOMAGENETIC CAPTURE OF ADENOVIRUSES

8.1 Introduction

Investigations of pathogenic viruses in the environment are challenged by the low concentration of viral contaminant present and the high level of nontarget interferences that co-purify with the target organism. Application of molecular biology-based technology in recent years has improved the sensitivity of detection for pathogens that present at low concentrations but has also raised questions regarding the infectivity of the pathogens that are detected by genome-based methods and potential human health risks. The rationales for developing immunomagnetic (IM) beads capturing for adenoviruses in this study are twofold: 1) to improve the current viral concentration/purification method for environmental samples, and 2) to separate intact viruses with capsid and surface antigens from free viral nucleic acid. The researchers hoped first to use antibodies specific to the adenovirus surface protein to capture the viral particles and concentrate the captured viral particles by the antibody-linked magnetic beads, followed by real-time PCR assay of the viral quantity. Because binding to the viral protein coat requires the viruses to be intact, theoretically IM capture ensures that most of the viruses captured are potentially infectious.

8.2 Materials and Methods

8.2.1 Adenovirus

The adenovirus used for the capture study was a recombinant adenovirus, rAd-5. The rAd-5 is adenovirus serotype 5 with a deletion of E1 (E1A + E1B) gene in its genome, which is replaced by a gene-encoding green fluorescence protein (GFP). The rAd-5 is identical to its native counterpart in morphology and physiology. Although rAd-5 can infect many different cell lines or tissues of human and nonhuman origin, the viral genome cannot replicate in the cell without E1 function. The propagation of rAd-5 is only possible within 293A cells (embryonic kidney cells) that constitutively express adenovirus E1 protein to complement the missing E1 function. This virus, with abortive infection, provides a safe system for a seeding study. With an inserted GFP gene in its genome, rAd-5 expresses GFP in infected cells. GFP can be visually and quantitatively detected by epifluorescence microscopy. The viral infectivity and titer can be defined within 24 to 48 hours in comparison with a traditional plaque assay that requires two to three weeks. In addition, rAd-5 can propagate to a very high titer required for seeding. In all experiments conducted in this study, rAD-5 stock was titered by plaque assay and diluted in PBS before being used for experiments.

8.2.2 Magnetic Beads and Antibodies

The magnetic beads used in this study were purchased from Dynabeads (Dynal A/S, Oslo, Norway). Two types of beads were used: Dynabeads M-280 Tosylactivated and M-280 beads precoated with sheep anti-mouse IgG. Dynabeads M-280 Tosylactivated beads are uniform, superparamagnetic, polystyrene beads coated with a polyurethane layer. The polyurethane surface is activated by p-toluenesulphonyl chloride to provide reactive groups for covalent binding of proteins (antibodies) or other ligands containing primary amino or sulphhydryl groups. Dynabeads M-280 (Tosylactivated) binds proteins physically and chemically with increasing amounts of covalent bonds with increasing temperature and/or pH. Dynabeads M-280 (Tosylactivated) is supplied as ethanol-washed beads in aqueous suspension and is designed as a solid phase for biomagnetic separations. They are intended for coating with antibodies or other selecting molecules. Any protein, peptide, or glycoprotein may be bound to its surface. Dynal Biotech recommends early blocking with an irrelevant protein such as BSA when antibodies or other ligands are coated directly onto the surface of the beads, to ensure optimal availability of the antibody binding sites. Beads precoated with sheep anti-mouse IgG were also used to capture the primary antibody in the case of mouse anti-adenovirus monoclone antibody. The structure of primary antibody to secondary antibody binding allows the correct position of anti-adenovirus capturing arms, therefore increasing the efficiency of target capturing.

Three different types of monoclonal antibody specific to adenoviruses were purchased from Fitzerald International Corp. (Concord, Ma.). Clone M0111148 (antibody A) and M0111149 (antibody B) are both specific for the hexon group antigen of adenovirus with reactivity with many serotypes. Clone M1121801 (antibody C) is specific to a broad spectrum of adenoviruses without specification of an antibody binding site. A polyclone goat anti-adenovirus from Fitzerald International Corp. was also purchased and tested.

8.2.3 Beads Preparation and Viral Capture

Dynabeads were washed twice with PBS buffer containing 0.1% of BSA at room temperature for 15 min each time with gentle rotating. Anti-adenovirus antibodies were linked to the beads at 4°C overnight for uncoated Dynabeads or at room temperature for one hour for sheep anti-mouse IgG precoated beads. Variations of incubation temperature and time were tested to improve capture efficiency. The beads to antibody ratio was also increased from 3:1 to 5:1 in an effort to increase the capture efficiency. Beads that were not coated with anti-adenovirus antibody were used as the control to evaluate antibody binding efficiency and the occurrence of nonspecific binding.

For viral capture, 1 ml of adenovirus stock diluted in PBS was added to antibody coated beads. The final concentration of adenovirus tested ranged from 1 to 10^5 PFU/ml. The mixture was incubated overnight at 4°C. The next morning beads were magnetized and washed with PBS with 0.1% BSA three to five times. Also tested was the effect of adding 0.1% SDS and 0.05% Tween 20 on removal of nonspecific binding during the washing step. The washing temperature was increased from room temperature to 37°C in attempts to remove nonspecific binding.

8.2.4 Efficiency of Antibody Binding to the Dynabeads

The efficiency of antibody binding to Dynabeads was assayed by Western Blot. In brief, protein on the beads was solublelized by boiling and separating by one dimension SDS denature gel, then transferred to the nylon membrane for western hybridization with HRP-labeled goat anti-mouse IgG. The hybridization was then exposed to X-ray film.

8.2.5 PCR Detection of Beads Captured Viruses

After the final wash, Dynabeads were resuspended in 25 µl of 1x PCR buffer. Then a 10-µl aliquot was heated to 95°C for 5 min to release viral nucleic acid, followed by the addition of

Ad-1/Ad-2 primers, dNTP, and Taq polymerase following the protocol developed by Xu et al. (2002). PCR was performed using a GeneAmp 2400 thermal cycler (Applied Biosystems) with 94°C for 5 min, followed by 30 cycles of denaturation at 94°C for 1 min, annealing at 54°C for 45 sec, extension at 72°C for 2 min, and a final extension at 72°C for 5 min. The amplicon was visualized on 1% agarose gels.

8.3 Results and Problem Encountered

Figure 8-1 shows the results of adenovirus rAd5 capturing using Dynabeads coated with different combinations of monoclonal antibodies and negative-antibody controls. It appears that all combinations of antibodies and no-antibody BSA controls yielded the similar results. The lowest concentration detectable by anti-adenovirus antibody coated beads was 100 PFU. Thus the efficiency of capture observed here is much lower than the previous report of immunomagnetic capture of hepatitis A viruses (0.1 PFU). In addition, the control beads that did not contain any anti-adenovirus antibodies also captured adenoviruses at 1000 PFU. Therefore, the results suggest that viruses captured by Dynabeads in these experiments were likely the result of nonspecific binding to the beads rather than to the viral-specific monoclonal antibodies.

To remove the nonspecific binding, the researchers experimented with several different washing conditions. Adding 1% SDS, 0.05% Tween 20 or increasing the washing temperature from room temperature to 37°C did not remove the nonspecific binding. Increasing the washing temperature to 45°C resulted in loss of all viral signals from both antibody-coated and uncoated beads (data not shown).

In an attempt to increase viral capture efficiency and removal of nonspecific binding from the beads, the researchers experimented with beads precoated with sheep anti-mouse IgG. It was hoped that physical position of the anti-adenovirus IgG capturing arm might improve the capturing efficiency. In addition, the lack of Tosylactivated surface would reduce or eliminate nonspecific binding. However, results as shown in Figure 8-2 were similar to those observed with M-280 Tosylactivated beads.

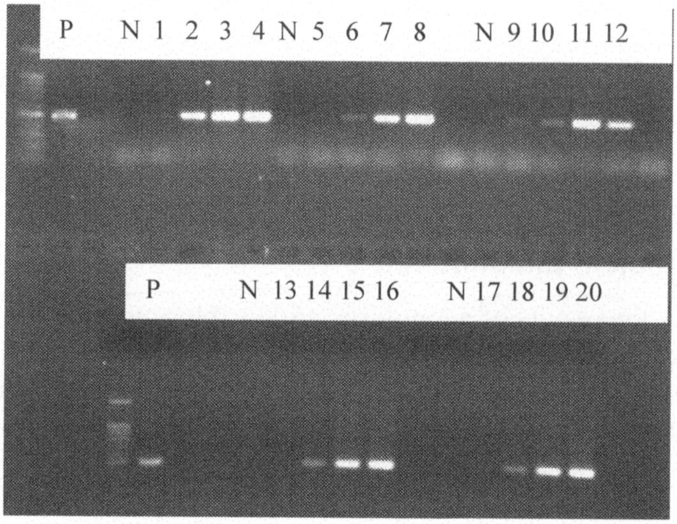

Figure 8-1. Testing the Efficiency of Immunomagnetic Capture of Adenoviruses with Different Combinations of Monoclone Antibodies.
Beads coated with antibody B & C were used in Lanes 1-4; with antibody A & C in Lane 5-8; with antibody A & B in Lanes 9-12; with antibody A & B & C in Lane 17-20. Beads used in Lanes 13-16 did not coat with antibody.
P: PCR Position Control; N: PCR Negataive Control; each type of bead was tested for capture efficiency/nonspecific binding using a series of dilution of adenoviruses from 10^2 to 10^5 PFU (from left to right).

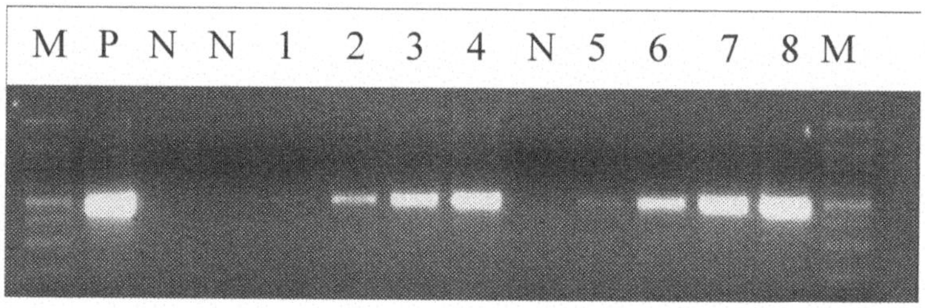

Figure 8-2. Testing the Immunomagnetic Capture of Adenoviruses Using Beads Precoated with Sheep Anti-mouse IgG and a Combination of Three Types of Monoclone Antibodies (A & B & C).
P: PCR Position Control; N: PCR Negataive Control. Beads with three anti-adenovirus monoclone antibodies combination were used in Lanes 5-8. Beads without anti-adenovirus antibody were used in Lane 1-4s as control. Each type of bead was tested for capture efficiency/nonspecific binding using a series of dilution of adenoviruses from 10 to 10^4 PFU (from left to right).

A possible explanation of the low efficiency of viral capture may have resulted from low binding of anti-adenovirus antibody to the beads. To rule out this possibility, the amount of anti-adenovirus antibody on coated beads was assayed by western blot (Figure 8-3). The results indicated that the concentration of antibody linked to the beads was equal or greater than 1.0 μg. Thus, the antibody binding was an unlikely reason for low capturing efficiency.

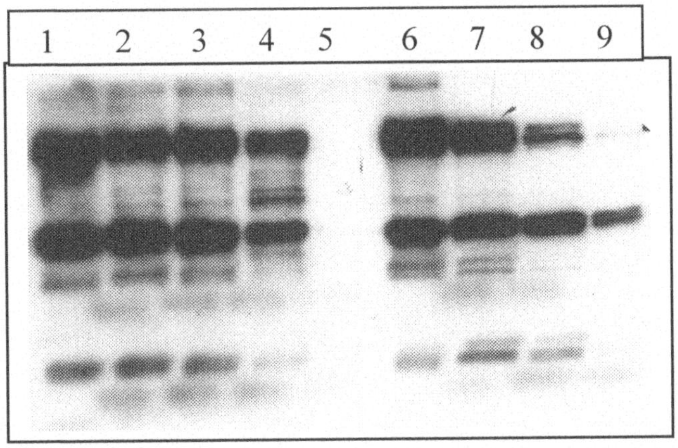

Figure 8-3. Assay of the Efficiency of Anti-adenovirus Antibody Binding to the Magnetic Beads.
Antibodies were eluted from the magnetic beads and separated on SDS gel. Western blot as probed with HRP-labeled goat-anti-mouse IgG. Lanes 1 to 6 are beads coated with antibody 1) A&B, 2) A&C, 3) B&C, 4) A&B&C; 5) no antibody; 6) sheep-anti-mouse IgG-A&B&C; Lanes 7-9 are dilutions of mouse IgG as quantification control 7) 1.0ug; 8) 0.5 ug; 9) 0.1 ug.

In a final attempt, a polyclonal antibody was evaluated. Similar results to those seen in Figure 8-1 were observed. Therefore, the nonspecific binding and low antibody capturing efficiency may not be due to the type and quality of the antibody.

Consultation with Project Subcommittee members resulted in a recommendation to experiment IMC with highly purified adenoviruses. Following this recommendation, a batch of rAd5 was produced and purified twice using cesium chloride gradient ultracentrifugation. The viral band was extracted and dialyzed to removed salt. When this highly purified rAd5 was applied to the capturing study, results similar to those in Figure 8-1 were once again observed, suggesting that additional study of the physical, chemical, and biological characters and immunological properties of adenoviruses are necessary before the problem in IMC can be resolved. Such studies were beyond the scope of the current project.

CHAPTER 9.0

QUANTIFICATION OF ENTEROCOCCI IN ENVIRONMENTAL SAMPLES BY REAL-TIME PCR

9.1 Introduction

Enterococcus is part of the normal flora of the gastrointestinal tract in humans and animals and has emerged as leading causes of nosocomial infection[61, 62]. There are two major pathogenic species in humans, *Enterococcus faecalis (E. faecalis)* and *E. faecium*, with occasional infections being caused by *E. durans, E. gallinarum, E. casseliflavus, E. avium, E. hirae, E. mundtii,* and *E. raffinosus*[63]. Enterococcal infections are especially troublesome because of the high level of intrinsic antibiotic resistance[64]. The prolonged viability of the naturally occurring *E. faecalis* under simulated deep-sea conditions is a concern with respect to the contamination of the deep ocean with human pathogens and the possibility of sewage-associated microorganisms. At the recreational beaches, the abundance of *Enterococcus* in bathing waters is correlated with the incidence of swimming-related gastroenteritis[29]. In order to protect human health during water recreation, *Enterococcus* has been used as an indicator of fecal contamination in the costal waters since 1999 (California AB411). Thus, the rapid detection of *Enterococcus* in the environment is important in reducing the spread of multiresistant *Enterococcus* and also in assessing the quality of bathing waters. However, the conventional methods for *Enterococcus* detection based on cultivation often require one to two days[65]. Additionally, growing bacteria in artificial media contributes to the poor cultivability of injured and stressed organisms. *Enterococcus* may also enter a viable but not culturable (VBNC) state[66], which may further limit the use of traditional culture methods.

PCR has become an important method for the rapid, sensitive, and specific detection of bacterial and viral agents. For example, it has been used qualitatively but not quantitatively for detection of *Enterococcus* in environments[19, 67]. Recently, real-time PCR was applied to quantitatively determine the presence of microorganisms in the complex medium. This method improves the accuracy and sensitivity of traditional single step PCR by adding a fluorescently labeled probe so that the target gene can be detected without subsequent verification. Among the various quantitative PCR strategies available, those based on real-time monitoring of the amplification reaction are the most accurate[68]. A 16S rDNA real-time PCR method was developed to detect *E. faecalis* in water samples[69]. The dynamic range for cell detection spanned 5 logs and the detection limit was determined to be 6 CFU/reaction without nucleic acid extraction, indicating that real-time PCR is very robust and sensitive in quantification of *Enterococcus* in environmental samples [69]. More recently, real-time PCR was applied to detect and quantify *E. coli* O157:H7 [70] in foods, *Enterococcus spp.* in drinking water [2] and monitor nitrifying bacteria in a municipal wastewater treatment plant[71].

In the present study, the development of a real-time PCR method for quantification of *Enterococcus* in environmental waters is described.

9.2 Materials and Methods

9.2.1 Bacteria and Culture Conditions

Enterococcus faecalis was purchased from American Type Culture Collection (ATCC) and was used for real-time PCR optimization, sensitivity assay, and creating quantification standards. The cell density of *Enterococcus* in culture and in environmental samples was enumerated using the membrane filtration method or enteroalert (Idexx Inc.) and counted as colony forming unit (CFU)/100 ml or most probable number (MPN)/100 ml of sample.

9.2.2 Environmental Samples

Sewage samples were all 24-h composite samples collected by random sampling in two Southern California sewage treatment plants. Primary effluent was raw domestic sewage after sedimentation. Secondary effluent was treated by active sludge. Some of the active sludge effluent was also treated with chlorine at various locations in the effluent transportation pipeline. The final concentration of chlorine residue was between 2-3ppm in the composites.

Coastal water samples were collected from a small estuary, Newport Bay, Calif. and its tributary, San Diego Creek. The salinity of the water samples was 1 ppt (parts per thousand) in San Diego Creek, approximately 15 ppt in the mid Newport Bay and near 30 ppt in the lower Newport Bay where bay water is well mixed with ocean water. In addition, 30 water samples were collected from various locations and depths from Newport Dunes, a designated recreational beach and R/V park, in the mid Newport Bay.

9.2.3 Quantification Standards for Real-time PCR

To create standards for *Enterococcus*, the whole bacterial genomic DNA of *E. faecalis* was extracted using QIAamp DNA Mini Kit and quantified by UV spectrophotometer. The concentration of bacterial genomic DNA was converted to genome equivalent and serially diluted to a range of 2×10^3 to 2×10^9 copies/ml: a 2.5-microliter of each dilution (5 to 5×10^6 copies/reaction) was used in triplicate as a quantitative standard for *Enterococcus*.

9.2.4 Sample Preparation for Real-time PCR

For real-time PCR of *Enterococcus*, 1 to 10 ml of *E. faecalis* cultures or environmental samples was centrifuged at 8,000 rpm for 5 min. The resultant bacterial pellet was resuspended in 100 μl of lysis buffer [10 mM Tris-HCl (pH 8.3), 100 mM NaCl, 1mM EDTA, 1% (v/v) Tween-20 and 10mg/ml of proteinase K] and incubated at 45°C for 3 hours. For coastal water samples, the lysates were extracted with 100 μl phenol/chloroform/isoamoalcohol (pH.8.0) and the supernatant was precipitated with ethanol. The resultant DNA pellet was washed with 70% ethanol and resuspended in 30 μl ddH_2O. For sewage samples, the upper liquid phase from phenol/chloroform extraction was further purified using QIAamp DNA Mini Kit (Qiagen) according to the manufacturer's instruction to remove PCR inhibitors and interferences. Purified bacterial DNA was eluted with 30 μl ddH_2O.

9.2.5 Real-time PCR

Real-time PCR for direct quantification of *Enterococcus* was developed based on the previous report of Frahm, et al[2]. The *Enterococcus* primers and TaqMan probe target 23S rRNA gene (Table 9-1).

Real-time PCR was carried out in 25-μl reaction mixtures consisting of 1x TaqMan master mix, primers, TaqMan probe, and template DNA. Concentration of primers and probe was optimized using concentration matrix to achieve the best amplification efficiency (Table 9-1). Samples were amplified and detected using ABI Prism 7000 Sequence Detection System (Applied Biosystems). The PCR amplification conditions were optimized following the manufacturer's recommendations (Applied Biosystems). The final thermocycling profile for *Enterococcus* real-time PCR was 95°C 15 sec, 60°C 1 min for 40 cycles. All samples were run in triplicates.

9.3 Results

9.3.1 Standard Curve and Sensitivity of Real-time PCR

To achieve the optimal real-time PCR settings that give reproducible results with a high sensitivity, various amplification conditions were tested as described in the previous chapters. In brief, these conditions include a variation of oligonucleotide concentration for each primer from 50 to 900 nM, a variation of probe concentration from 50 to 250 nM, a variation of thermocycling profile according to the primers and probe Tm, and an optimization of cycle length and numbers (data not shown). The final optimal PCR conditions are presented in Table 9-1.

Figure 9-1 shows the plot of real-time PCR amplification of *E. faecalis* using known copies of genomic equivalent as a standard and a serial diluted bacterial culture as a sensitivity calibrator. The results showed a positive log linear relationship of *E. faecalis* genome copies and PCR threshold cycle numbers, expanding a 7-log unit, between 5 to 5 $x10^6$ /reaction. The correlation coefficient of the standard curve was 0.99, while the slope was 3.3. The sensitivity of detection for serial diluted known numbers of *E. faecalis* suspension counted by culture was comparable to the standard curve. The lower sensitivity limit was near 5 CPU/reaction (Figure 9-1). Compared to the standard curve, however, a lower slope of the diluted bacteria was noted at the lower concentration range (5 to 50 CFU/ml). This discrepancy is likely due to the loss of template during DNA preparation from the diluted bacterial samples.

Table 9-1. Summary of Primers, Probe, and Conditions for Detection of *Enterococcus*.

Name	Sequence	Conc.(nM)	Tm (°C)	Assay Targets
Enterococcus				
ECF	AGAAATTCCAAACGAACTTG	900	53.1	*faecalis, faecium,*
ECR	CAGTGCTCTACCTCCATCATT	300	5.4	*durans, asseliflavus,*
ECP	FAM-TGGTTCTCTCCGAAATAGCTT TAGGGCTA-TAMRA	250	67.2	*gallinarum and hirae**

FAM, 6-carboxyfluorescein, fluorescence reporter dye.
TAMRA, 6-carboxy-tetramethylrhodamine, fluorescence quencher dye.
* summarized from publication [2].

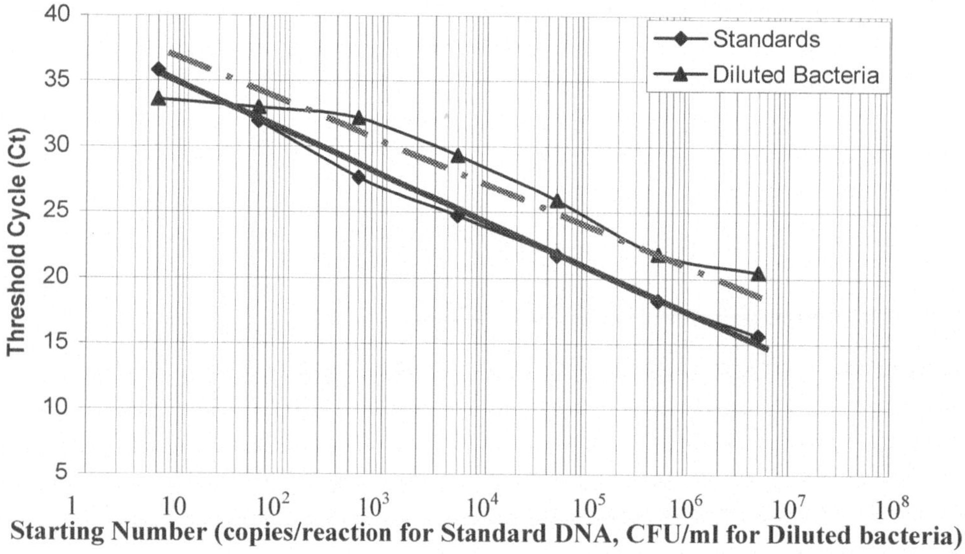

Figure 9-1. Standard Curve and Sensitivity of Real-time PCR for Detection of *Enterococcus*.
Genomic DNA (♦) extracted directly from *Enterococcus faecalis* with known number of genome copies, was wsed as the standard. Serially diluted *Enterococcus faecalis* culture (▲) that was enumerated by plate counts on LB medium was used as the sensitivity calibrator. Each data point represents the threshold cycle (Ct) average of samples prepared in triplicate. Blue solid line is the standard curve and red dash line is the sensitivity curve.

9.3.2 Application of Real-time PCR for Quantification of *Enterococcus* in Environmental Waters

To address the feasibility of using real-time PCR to detect *Enterococcus* in environmental waters, the researchers applied this method to various environmental water samples including sewage, brackish coastal waters, and creek samples and compared the results with the traditional culture method (Table 9-2). The first few attempts at direct quantification of *Enterococcus* in environmental waters failed to yield satisfactory results due to the sensitivity of real-time PCR assay to inhibitors and interferences in the environmental samples (data not shown). Further improvement of *Enterococcus* lysis and DNA purification protocols have proven successful (see Materials and Methods). A QIAamp DNA Mini column purification step was necessary to further remove PCR interferences and inhibitors from sewage samples. This extra step made the real-time PCR method feasible for sewage samples (data not shown).

Table 9-2 shows the comparison of *Enterococcus* quantification by real-time PCR and culture assay in environmental waters. The results demonstrate that real-time PCR method was applicable for direct quantification of *Enterococcus* in primary, secondary, and chlorinated secondary sewage effluents after sample purification. The bacterial concentration determined by quantification PCR ranged from 10^5 to 10^6 genomic copies/100 ml in sewage samples, which is comparable to the *Enterococcus* counts determined by culture assay in general. However, the real-time PCR method significantly overestimated bacterial concentrations in chlorinated secondary effluents, suggesting that chlorination injured or inactived the cells but did not complete degrade the target DNA. When the real-time PCR was applied to creek and coastal waters, this method also overestimated the bacterial concentrations, as shown in Table 9-2, by one to two orders of magnitude, especially for samples collected from saline environments such as the mid and lower Newport Bay. Of 31 samples collected from various locations from recreational beaches of Newport Dunes, only four samples gave low signals for real-time PCR, while all samples were below culture assay detection limit for *Enterococcus*.

Table 9-2. Comparison of *Enterococcus* Quantification by Real-time PCR and Culture Assay in Environmental Waters.

Sample	Real-time PCR (Genomic copies/100 ml)	Culture assay* (CFU or MPN/100 ml)
Sewage		
9/15/03		
Primary effluent–Plant I	$3.5x10^5$	$8.1x10^5$
Secondary effluent–Plant I	$1.4x10^5$	$6.0x10^5$
Secondary effluent–Plant II	$1.7x10^5$	$3.2x10^4$
9/30/03		
Primary effluent–Plant I	$2.6x10^6$	$1.8x10^6$
Secondary effluent–Plant I	$3.5x10^5$	$1.4x10^5$
Primary effluent–Plant II	$1.6x10^5$	$1.7x10^5$
Secondary effluent with chlorine–Plant II	$6.9x10^5$	0
10/1/03		
Primary effluent–Plant I	$6.4x10^5$	$2.0x10^6$
Secondary effluent–Plant I	$2.1x10^5$	$2.3x10^4$
Primary effluent–Plant II	$8.1x10^5$	$4.2x10^5$
Secondary effluent with chlorine–Plant II	$5.4x10^5$	0
Coastal Waters		
San Diego Creek	$4.5x10^3$	$8.8x10^2$
Mid Newport Bay	$2.4x10^3$	7.2
Lower Newport Bay	$7.3x10^2$	59
Newport Dunes (multiple sites n=4)	$(2.8\pm2.0) \times 10^2$	Below detection**
Newport Dunes (multiple sites n=27)	Below detection	Below detection

*All samples except Newport Dunes samples were determined by membrane filtration; Newport Dunes samples were determined using Idexx kit.
**Detection limit for real-time PCR is 5 CFU/reaction, for culturing assay is 10 MPN/100 ml

9.4 Discussion

In this study, the researchers demonstrated that the real-time PCR method for *Enterococcus* allowed quantification of this organism in environmental waters. They chose *E. faecalis* whole genomic DNA instead of plasmid DNA with a single insert of target sequence as a quantification standard, as there are multiple operons of 23S rRNA gene in each *Enterococcus* genome. Using recombinant plasmid DNA with a single insert of 23 rRNA gene as a standard yields the genome equivalence of 23S rRNA gene in samples but may not be accurately converted to the number of CFU of *Enterococcus* in environmental samples, thus complicating the interpretation of the quantity of *Enterococcus*. The primers and probe for *Enterococcus* real-time PCR used in this study was originally described by Frahm et al[2]. However, this method was not applied quantitatively to the environmental samples. In Frahm et al. an incubation step was required to enrich the bacteria. The enrichment itself tarnished the characteristic of real-time quantification. In current the study, the researchers optimized the ratio of primers to probes from the previously reported 200, 300/360 nM [2]to 900, 300/250 nM to achieve approximately twofold increases in signal intensity (results not shown). They also improved the extraction and purification method to achieve optimal bacterial DNA purification and removal of PCR inhibitors in water samples. These improvements have allowed this method to be used in environmental samples for quantification of *Enterococcus* in less than 4 hours. In comparison with the quantitative real-time PCR method reported by Santo Domingo, et al.[69], the current method is targeting a broad range of *Enterococcus* spp. Rather than a few specific species, therefore, the results are more comparable to environmental samples assayed using cultured methods.

The discrepancy between the real-time PCR quantification and *Enterococcus* culture counts was observed in the chlorinated sewage effluents and coastal waters. This is not surprising because it is well known that chlorination and physical stress may injure the cell but not completely destroy the bacterial genome. Real-time PCR may also detect naked nucleic acid, which has little representation of cultured *Enterococcus*. In addition, the VBNC state may also account for the discrepancy. VBNC is a survival mechanism of bacteria facing environmental stress conditions[72]. Bacteria are no longer able to grow and form colonies on conventional culture media but demonstrate metabolic activity, maintain their pathogenicity, and, in some cases, may return to active growth when optimal conditions are restored[73]. Recently, *E. faecalis* has been demonstrated to be able to enter the VBNC state[66]. Furthermore, an RT-PCR has been established to detect *E. faecalis* pbp5 mRNA to monitor the viability of VBNC[66]. VBNC lends further support to criticisms of the traditional methods used to evaluate water quality based on plate counts. *Enterococcus* detected by real-time PCR in coastal waters is likely to be in the VBNC state because several reports have suggested that many types of bacteria enter this state in nutrient poor environment[72]. Additional work using a culture resuscitation method or VBNC counts may provide an answer to this hypothesis.

In summary, this study demonstrated the application of real-time PCR for quantification of enterococci in environmental waters. Further improvements of DNA extraction and purification methods are likely to improve the accuracy of these quantification methods. However, because PCR detection reflects little of cultured bacteria, the outcome of the real-time result is inadequate to represent the efficacy of sewage treatment process.

CHAPTER 10.0

REAL-TIME PCR QUANTIFICATION OF HUMAN ADENOVIRUSES IN URBAN RIVERS

10.1 Summary

Real-time PCR was applied to quantify the abundance of human adenoviruses in two Southern California urban rivers, San Gabriel and Los Angeles. A total of 114 river samples from five different locations were collected over a one-year period and analyzed for human adenoviruses along with fecal indicator bacteria and coliphages. Adenoviruses were detected by real-time PCR in approximately 16% of the samples, with concentrations ranging from 10^3 to up to10^4 genomes per liter. However, plaque assay using two human tissue culture cell lines, HEK-293A and A549, yielded negative results, suggesting adenoviruses detected by real-time PCR are likely noninfectious. Enterovirus was detected in ~7% of the samples by reverse-transcriptional PCR. No statistical correlation between human adenoviruses and fecal indicator bacteria (total coliform, fecal coliform, and *Enterococcus*) was observed. This chapter presents the first quantitative measurement of human adenoviruses in urban rivers and their statistical relationship to fecal indicator bacteria and coliphages.

10.2 Introduction

Rapid population growth and urbanization occurring along coastal regions has made Southern California one of the most densely populated areas in the U.S.. More than two-thirds of Californians visit local beaches at least once a year. Southern California beaches have become one of the most desired beach vacation spots in the world, attracting over 100 million visitors to the shoreline annually and offering recreational activities such as swimming, diving, yachting, surfing, snorkeling, windsurfing, fishing, hiking, and cycling. Correspondingly, an increase in the volume of domestic sewage discharges and urban runoff to the coastal ocean has become a major public health concern in Southern California[5, 74]. In 2001, a total of 1,592 Orange County beach closures and advisories were noted as a result of intolerable contamination levels, an 81% increase from the previous year (http://www.healthebay.org/).

Microbial contamination of the aquatic environments poses a potential public health risk when improperly managed[75, 76]. The U.S. EPA has established microbial pollution indicator standards recommending the use of bacterial indicators, including total and fecal coliforms and enterococci, during routine monitoring[3, 4, 29]. However, recent studies show that current bacterial indicator standards may not accurately indicate the viral quality of waters[77]. A study conducted by Jiang et al. [5] found over 30% of coastal waters tested in Southern California contained human viral signals, with presence of viruses showing no association with elevated levels of the bacterial indicators.

Proper monitoring of human viruses in waters is of particular importance because the Centers for Disease Control and Prevention suggests that the causative agent of nearly 50% of all acute gastrointestinal illnesses are suspected to be viral[6]. However, current clinical diagnostic methods are insufficient for monitoring viral diseases. With approximately 100 potentially water transmissible human viruses associated with human waste, it is impossible to detect all viruses[7]. However, certain human viruses and coliphages may beneficially serve as an index in determining viral contamination and the presence of human fecal contamination[5, 18, 78]. Adenovirus, a double stranded DNA virus belonging to the family adenoviridae, is a possible candidate as an index virus. Frequently found in urban rivers and polluted coastal waters[5, 15-19], studies conducted in Europe suggest that adenovirus be used as an index of human viral pollution because this virus was frequently detected in human fecal contaminated samples[18]. Adenoviruses have also been shown to display greater resistance to UV-treatments because they contain double-stranded DNA, which allows for repair of damaged DNA using the enzymes of host cells[79, 80]. Thurston-Enriquez et al.[81, 82] showed that a UV dose of 103 mJ/cm^2 was required in order to achieve a 99% inactivation of adenovirus serotype 40 (1 of 51 serotypes of adenovirus) in treated groundwater. Currently, the UV doses commonly applied for drinking water and wastewater treatment is between 30 to 40 mJ/cm^2.

This study was conducted to quantitatively determine the seasonal distribution of viral and bacterial pollution in two of Southern California's urban rivers to determine their impact on coastal waters. This study also provides insight into the adequacy of current monitoring standards by determining if significant correlations exist between fecal indicator bacteria, coliphages, and human viruses (adenovirus and enterovirus).

10.3 Materials and Methods

10.3.1 Sampling Sites

Between February 2002 and February 2003, water samples were collected from three different sampling sites along the Los Angeles River and two sites along the San Gabriel River (Figure 10-1). The five sampling sites are designated LA1, LA2, and LA3 for the Los Angeles River, and SG1 and SG2 for the San Gabriel River. Sites SG1 and LA1 are located at the mouth of each river, and LA2, LA3, and SG2 are located upstream approximately 5-20 miles apart. All sites are located in urban settings; the three upstream sites are adjacent to major highways, while the two sites located at the mouth of the rivers are located in commercial settings. The water flow was fairly consistent for all sites at the Los Angeles River, while water level and water turbidity fluctuated in the San Gabriel River throughout the study period. In addition, recreational fishing activities were observed at both sites in the San Gabriel River. All sites in each river were sampled semi-monthly using 10-L containers sterilized with bleach. The samples were transported back to the laboratory at U.C. Irvine within two hours of collection for immediate processing. Sample salinity was determined by refractometry, water temperature was measured on site, and rainfall events were recorded. During the study period, a total of 114 10-L samples was collected from both rivers.

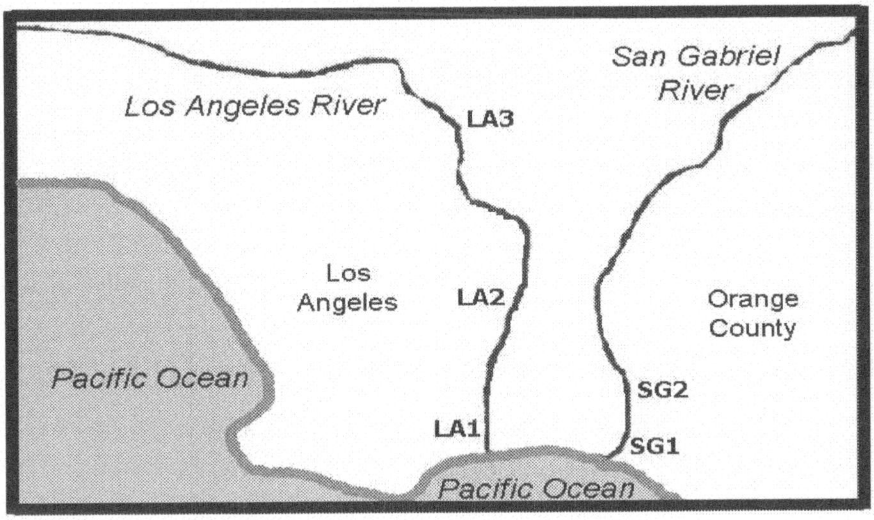

Figure 10-1. Sampling Locations along Los Angeles River and San Gabriel River.

10.3.2 Detection of Indicator Bacteria

Analysis of fecal indicator bacteria, total and fecal coliform, and *Enterococcus* was performed immediately after sample collection following the U.S. EPA's guidelines for bacterial indicator analysis by membrane filtration (Dufour et al., 1984, U.S. EPA, 1986 and 1994). In brief, environmental samples were filtered through 0.45-µm pore size nitrocellulose membranes. Fecal coliform and total coliforms on the membrane were cultured using m-FC agar (Fisher Scientific Inc.) and m-Endo agar (Fisher Scientific Inc.) and incubated overnight at 44.5°C and 37°C, respectively. For *Entercoccus* analysis, mE agar (Fisher Scientific Inc.) was used and incubated for two days at 41.5°C. Membranes were then transferred to EIA plates for 20 min for confirmation and recorded as CFU/100 mL.

10.3.3 Concentration of Water Samples for Viral Analysis

Ten-liter samples were concentrated to ~115 ml using a Tangential Flow Filtration system with a 30KD Omega cartridge at a flow rate of 350-450 mL/min under a constant pressure at <10 psi (TFF, Pall Life Science). The samples were spiked with bacteriophage ϕHSIC and were titered in samples prior to and after TFF concentration by plaque assay to determine the rate of viral recovery. ϕHSIC was isolated from coastal waters off Hawaii and has not been found in urban river waters of Southern California. Therefore, there are no interfering effects of viral recovery assay[83]. The head size of the phage is ~120 nm and the genome of the phage has been completely sequenced[84]. Immediately after concentration, 1 ml of the concentrated samples was used for coliphage plaque assay, and the rest were stored at -80°C for later use.

10.3.4 Detection of Coliphages

Densities of general and F-specific coliphage were determined by using both TFF-concentrated and unconcentrated water samples. Both 100 µL and 1 ml of the concentrated and unconcentrated samples were mixed with 1 ml of bacterial host in 3 ml of soft agar and poured over a LB agar plate. *E. coli ATCC* 15597 strain was used as the host for general coliphage, while

E. coli Famp was the specific host used for F-specific coliphage. Plaques were enumerated after 12-24 hours of incubation at 37°C.

10.3.5 Viral Nucleic Acid Purification by GuSCN-Silica Bead Extraction

The method of Boom et al [85] was used for nucleic acid purification and removal of PCR inhibitors for all environmental concentrates. Briefly, 900 µL of guanidinium thiocyanate lysis buffer (GuSCN) was added to a mixture of 40 µL of silica particle and 50 µL or 100 µL of viral concentrate at room temperature for 10 min. The mixture was then centrifuged, supernatant discarded, and pellet washed. The pellet was dried briefly and nucleic acid eluted using 50 µL of 1 X Tris–EDTA at 56°C.

10.3.6 Detection of Human Enterovirus by RT-PCR and Probe/Hybridization

The enterovirus-specific primer sequences used for reverse transcription-PCR (RT-PCR) were 5'-CCTCCGGCCCCTGAATG-3' and 5'- ACCGGATGGCCAATCCAA-3'. This primer pair amplifies a target at the highly conserved 5' untranslated region. The PCR protocol followed those previously described by Tsai et al. [55]. Briefly, reverse transcription was performed at 25°C for 10 min, followed by 42°C for 30 min and enzyme inactivation at 99°C for 5 min. The PCR amplifications were performed at 95°C for 2 min, followed by 30 cycles of 95°C for 1 min, 55°C for 1 min, and 72°C for 1 min, a final extension of 72°C for 10 min, and a hold step at 4°C. The PCR products were then analyzed on a 2% agarose gel.

To confirm the amplicon, DNA in gel was transferred to positively charged MagnaGraph 0.45 Transfer nylon membrane (Osmonics Inc.) via Southern Transfer. In brief, the gel was washed in 0.5 M NaOH/1.5 M NaCl and 1.5 M NaCl/0.5 M Tris (pH8) twice with each solution for 20 min. The blot was set up using 3MM Whatman paper cut to the size of the gel and soaked overnight in 10 X SCC (pH7). The membrane was then dried and the DNA covalently linked to the membrane by UV-crosslinking at 1200 X 100 $\mu J/cm^2$ twice.

After DNA transfer, nylon membranes were washed three times in 3 X SSC containing 0.1% SDS at room temperature for 10 min and at 60°C for 1 hour. The membranes were prehybridized for 1 hour at 37°C in solution containing 6x SSC, 5x Denhardt, 0.05% sodium phyrophosphate, 100 mg/ml of salmon sperm, and 0.5% SDS. Hybridization was carried out overnight at 45°C in solution containing 6x SSC, 1x Denhardt solution, 100 mg/ml of salmon sperm, 0.05% sodium pyrophosphate, and 25 ng/ml of the γ–[32^P] ATP labeled oligonucleotide probe. The sequence for the enterovirus probe is '5-TACTTTGGGTGTCCGTGTTTC3'. After hybridization, the membrane was washed three times at room temperature for 15 min and two more times at 55°C for 1 hour each time with a solution containing 6x SSC, 1% SDS, and 0.05% sodium pyrophosphate to remove any unbound probe. The membrane was sealed in a plastic bag and exposed to X-ray film for 24-48 hours at -80°C. A positive result was characterized by a signal on the film.

10.3.7 Detection of Human Adenovirus by Real-time PCR

For adenovirus analysis by real-time PCR, samples were further concentrated by ultracentrifugation. In brief, 10 ml of TFF concentrate was centrifuged at 41,000 rpm using the Beckman rotor SW41 for 1.5 hours at 10°C. The pellets were resuspended in ~200 µL of the supernatant and 100 µL was purified by GuSCN-Silica bead extraction (as described above). Either 1 µL or 4 µL of purified viral concentrate (TFF-ultracentrifugation) was used in triplicate reactions for real-time PCR. The standard curve used for quantitation was plotted from quadruplicate samples using C_t values of sixfold dilution of template extracted from cloned adenovirus serotype 5. A standard curve was run for each set of assay.

Each real-time PCR reaction (25 µL) contained 1x TaqMan Universal PCR Master Mix, 300nM of each primer, and 200nM TaqMan probe. Degenerative primers specific to human adenoviruses used were 5`-GACTCYTCWGTSAGYTGGCC-3` AND 5`-CCCTGGTAKCCRATRTTGTA-3`, and probe FAM-AACCAGTCYTTGGTCATGTTRCATTG~TAMRA as described by in Chapter 7.0. The PCR cycle conditions consist of holding samples at 50°C for 2 min and then 95°C for 10 min, followed by 45 cycles of 95°C for 15 sec, 56°C for 15 sec, and 62°C for 30 sec with 1 sec increments between each cycle. An Applied Biosystems 7000 Sequence Detector System was used for real-time PCR, and the data was analyzed with the ABI Prism 7000 software.

10.3.8 Tissue Culture Analysis for Infectious Human Adenoviruses

The infectivity of adenoviruses in environmental concentrates was analyzed using two cell lines, A549 and HEK-293A. The A549 human lung carcinoma cell line is currently the most commonly used cell line for adenovirus propagation and plaque titration, while the 293 human kidney embryonic cells are mostly used as a package cell line for production of nonreplicative adenovirus vectors[86-88]. HEK-293A is a recombinant cell line that continuously expresses adenovirus E1 protein for greater adenovirus replication[89], and 293A differs from original cell line 239 by its ability to adhere to the surface of the tissue culture petridish. After TFF concentration and ultracentrifugation, approximately 200µL of viral concentrate was purified by chloroform extraction, filtered through 0.2 µm low protein binding and retention Whatman filters (Whatman International Ltd), and overlaid onto monolayers of HEK-293A and A549 cells. After gentle shaking for 60 min at 37°C, the monolayers were washed with pre-warm PBS, and overlaid with 1.25% agarose containing 5% DMEM, 50 µg/ml gentamicin, and 2.5 µg/ml amphotericin B. The cell culture was incubated for seven to eight days in 5% CO_2 at 37°C, then a second overlay was carried out as described above, and incubated for an additional seven to 10 days. Cell cultures were examined under an Olympus Microscope using a 10x magnification lens for plaque formation.

10.3.9 Statistical Analysis

Spearman's rho correlation was used to determine correlations in seasonal distribution of fecal indicator bacteria, coliphages, and human viruses. Kruskal Wallis test was used to determine significant variability in parameters during rainy and nonrainy seasons. The SPSS program (SPSS Inc., IL) was used for all statistical analyses, and correlations were considered significant at a 95% confidence level.

10.4 Results

10.4.1 Indicator Bacteria and Coliphages

The concentrations of fecal indicator bacteria and coliphage are log-normal distributed throughout the study period and sampling location. To determine the temporal and spatial variability, distribution of fecal indicator bacteria and coliphages are reported in geometric mean value for each month or each location. All three fecal indicator bacteria were highest in concentration in November during a period of heavy rainfall (Figure 10-2a,b, and c). Based on the California recreational water quality daily limits (10,000 CFU/100 ml for total coliforms, 400 CFU/100 ml for fecal coliforms, and 104 CFU/100 ml for *Enterococcus* (California Assembly

Bill 411), 28% to 52.5% of the samples tested exceeded at least one daily bacterial indicator limit. Higher concentrations of fecal bacterial contamination were detected in LA2, LA3, and SG2, while bacterial contamination was approximately one order of magnitude lower at site SG1, located at the mouth of the San Gabriel River, compared to the other sites (Figure 10-2d).

The seasonal distribution of coliphage was similar to fecal indicator bacteria, with a major peak occurring within 48 hours of a rain storm during November (Figure 10-3a and b). Increases in coliphage concentration surpassing $1x10^2$ PFU/100 ml were also observed during the months of February, March, and June, while the lowest concentration was observed during the summer month of July. Concentrations of F-specific coliphage were generally consistent during the winter months, with the lowest concentration observed during the month of July, similar to somatic coliphage. Higher concentrations of coliphage were detected in the upstream sites of both rivers, similar to the distribution patterns of fecal indicator bacteria (Figure 10-3c). The highest concentrations for both F-specific and somatic coliphages were observed in LA2. Statistical analysis by Spearman's rho test showed significant correlations in seasonal distribution between coliphage and three fecal indicator bacteria ($p < 0.05$). Statistical analysis by the Kruskal Wallis test also showed a significant difference in coliphages and all three fecal indicator bacteria concentrations during rainy/nonrainy periods ($p = 0.053$).

10.4.2 Human Entero and Adenoviruses

Adenoviruses were detected in approximately 30% of all samples analyzed by real-time PCR. Higher frequency and median concentrations of adenovirus were observed during the winter months (Figure 10-4a). The concentration of adenoviruses ranged from 10^3 to up to 10^5 genomes per liter for a single sample and does not fit a log-normal distribution. The viruses were detected at all sampling locations except SG2 (Figure 10-4b). Statistical analysis using the Spearman's rho showed no clear relationship between adenovirus and other fecal indicators. Analysis by Kruskal Wallis test also showed no significant difference in adenovirus concentrations during rainy versus nonrainy periods.

Approximately 7% of all samples analyzed were positive for enterovirus by RT-PCR confirmed by hybridization. The seasonal distribution of enterovirus was sporadic, and no trend or correlation with any other parameters measured could be established as for adenovirus (Figure 10-5a). Enterovirus was slightly more prevalent in the Los Angeles River, with approximately 8.33% of the samples being positive, compared to the San Gabriel River, where approximately 5.88% of the samples were positive. However, enterovirus was detected in at least one environmental sample collected from each site during the study period (Figure 10-5b).

Results from the infectivity assay by two different cell cultures showed no positive results in all of the 114 environmental samples analyzed. A few samples were randomly selected and retested using a 2nd round of infection to ensure that the results were true negative. Cytotoxicity was observed in less than 3% of samples tested after purification.

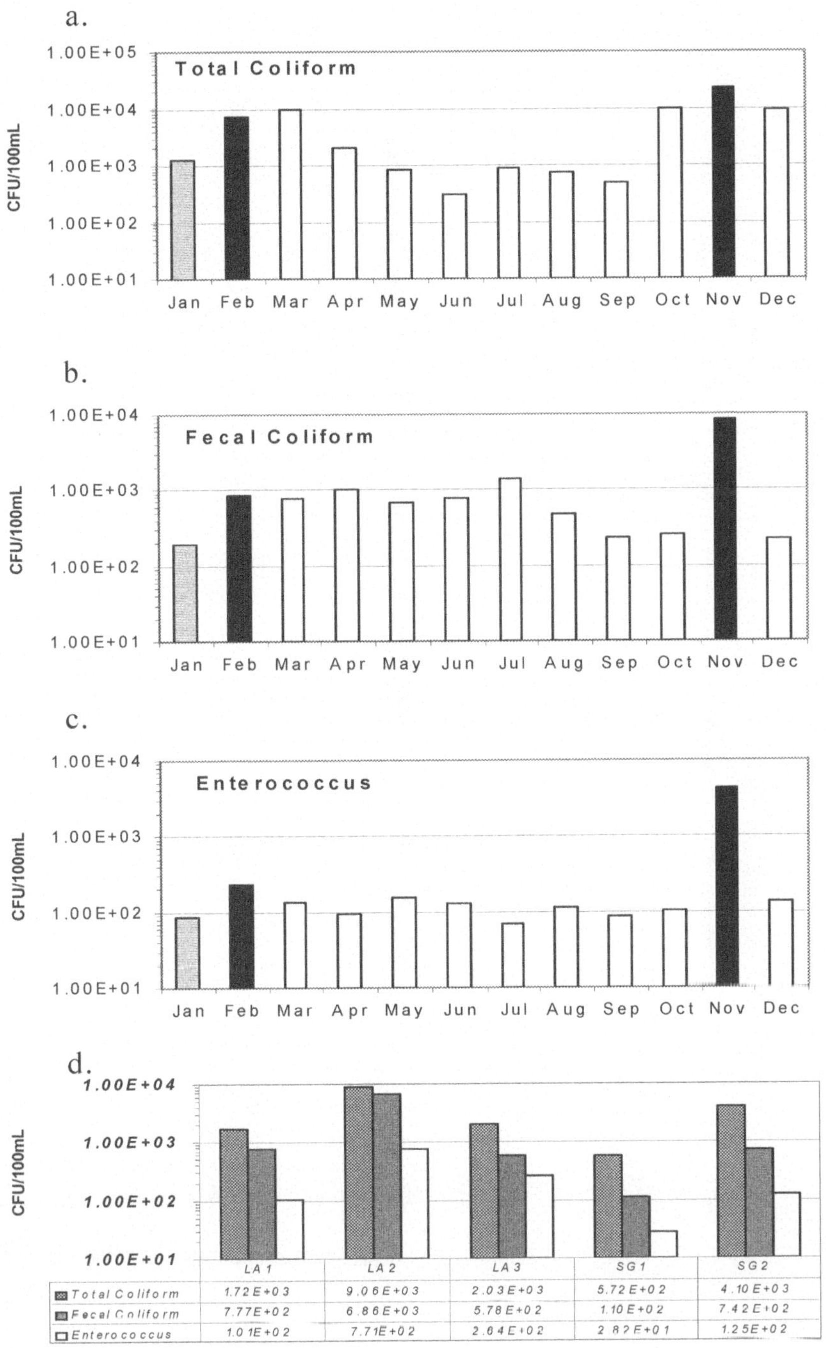

	LA 1	LA 2	LA 3	SG 1	SG 2
Total Coliform	1.72E+03	9.06E+03	2.03E+03	5.72E+02	4.10E+03
Fecal Coliform	7.77E+02	6.86E+03	5.78E+02	1.10E+02	7.42E+02
Enterococcus	1.01E+02	7.71E+02	2.64E+02	2.82E+01	1.25E+02

Figure 10-2. Seasonal Distribution of (a) Total Coliform, (b) Fecal Coliform, and (c) *Enterococcus* and Their Annual Genometric Mean Distribution at Each Sampling Location (d).

Storm events (■) were observed in February and November and light rain (□) was recorded in January within 48 hours of sampling. All bacterial indicators were measured using membrane filtration method following U.S. EPA protocols for each water sample.

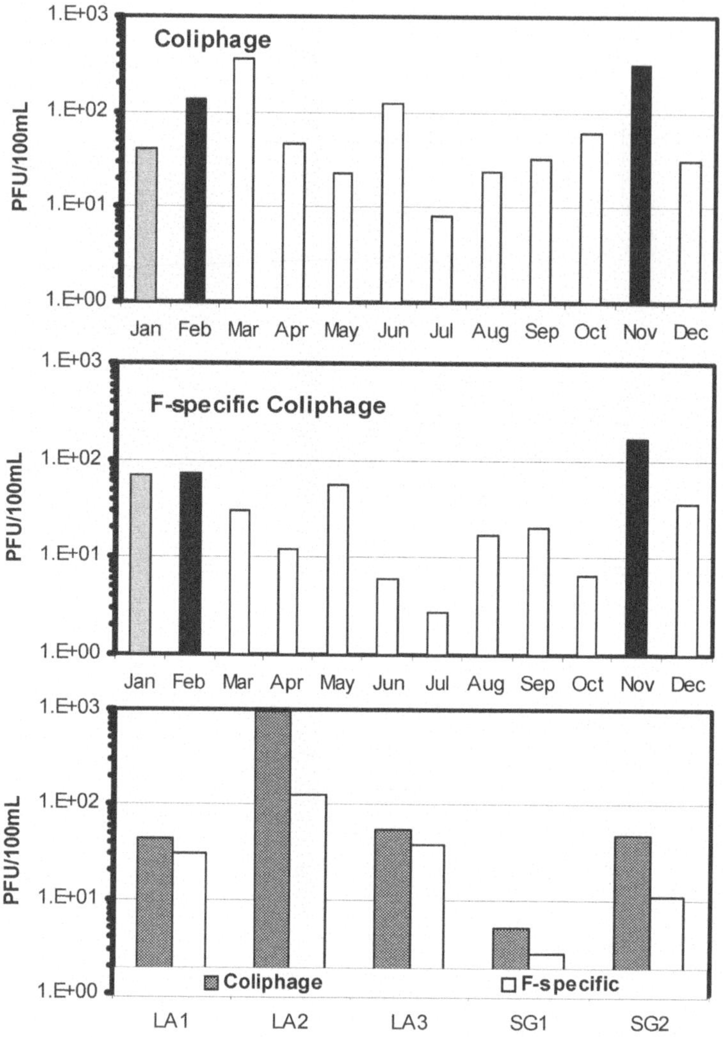

Figure 10-3. Seasonal Distribution of (a) Coliphage and (b) F-specific Coliphage and Their Annual Geometric Mean Distribution at Each Sampling Site (c).
Storm events (■) were observed in February and November and light rain (□) was recorded in January within 48 hours of sampling. Somatic and F-specific coliphage were measured using top agar overlay methods following U.S. EPA standard protocols.

a.

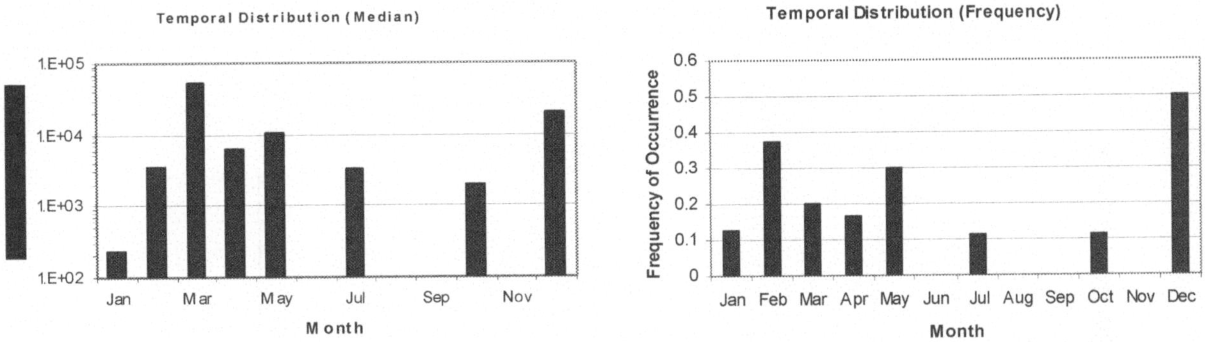

b

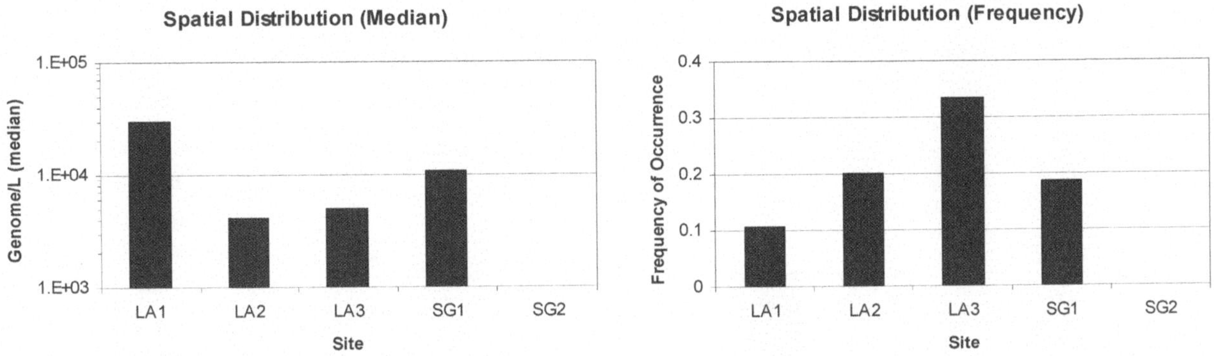

Figure 10-4. Seasonal Distribution of Adenoviruses (a) and their Median Concentration and Frequency of Detection at each Sampling Location (b).

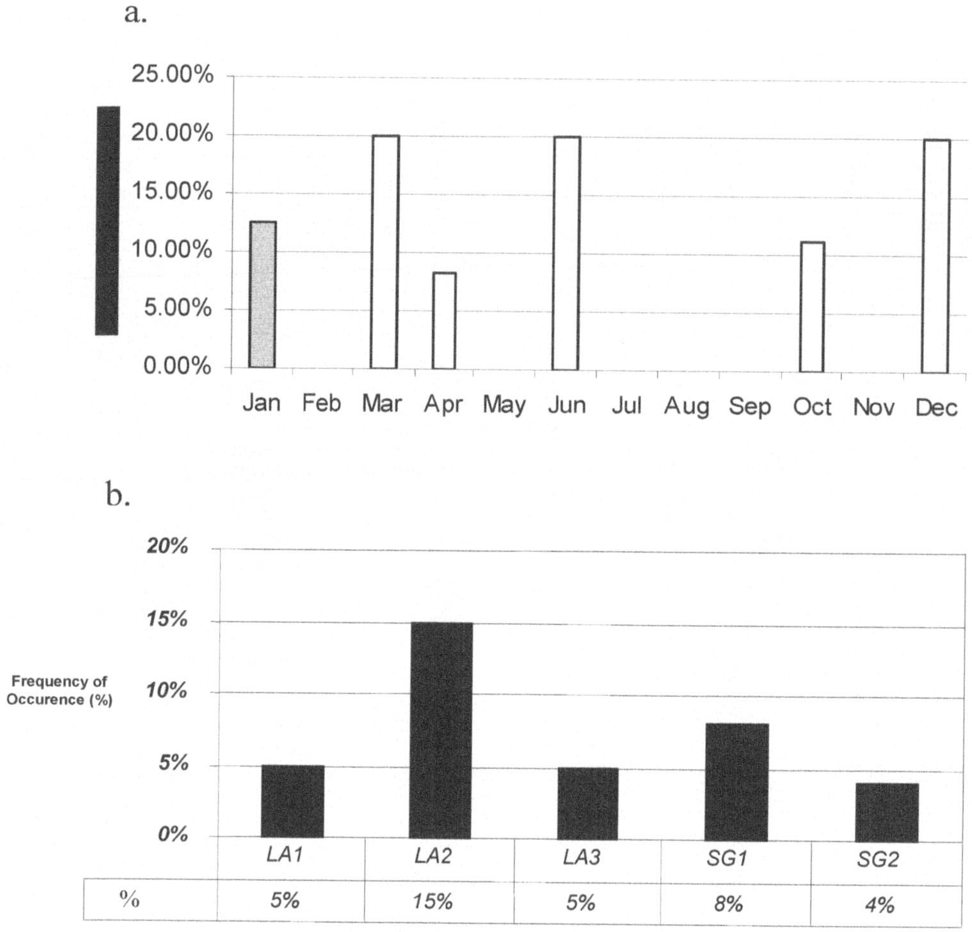

Figure 10-5. Seasonal Distribution of Enteroviruses (a) and their Frequency of Detection at Each Sampling Location (b). Storm events (■) were observed in February and November and light rain (□) was recorded in January within 48 hours of sampling. The presence and absence of enteroviruses were determined using RT-PCR followed by southern hybridization using an internal oligoprobe.

10.5 Discussion

Stormwater and urban runoff are increasingly being considered as major sources of coastal water pollution. Health concerns for beachgoers are clear, as numerous studies suggest that Southern California shorelines receiving runoff fail to meet California water quality standards. Noble et al.[90] showed that the failure to meet water quality standards increased 90% for shoreline areas adjacent to urban runoff outlets along the Southern California Bight shoreline following rainstorms. Similarly, Lipp et al.[91] demonstrated that the highest concentrations of fecal indicator bacteria were observed near areas where estuaries receive urban streams in Charlotte Harbor, Florida. Most recently, Jiang and Chu[92] showed that human virus contamination was widespread in Southern California coastal rivers and streams. This current study further investigated the presence of human pollution in urban rivers by presenting the first quantitative assessment of human viruses in urban rivers by real-time PCR and exploring their statistical relationships to fecal indicator bacteria and coliphages.

In accordance with previous observations, the current investigation shows that the Los Angeles and San Gabriel Rivers are contaminated with cultured fecal indicator bacteria and PCR-detectable human viruses. The concentrations of both fecal indicator bacteria and coliphages were higher at the upstream locations of both rivers, while high concentrations of adenoviruses were observed at both the mouth and upstream sites. A greater abundance of adenovirus genomes was detected at the mouth of each river in comparison to fecal indicator bacteria and coliphage, possibly because samples from upstream sites have a prevalence of humic material which results in a greater problem with PCR inhibition. One possible explanation for the discrepancy between the real-time PCR and tissue culture results is that natural or treatment processes such as sunlight or chlorination may have inactivated the adenoviruses while the genomes were undamaged, thus persisting in the environment for much longer periods of time and resulting in positive results by real-time PCR. Sobsey et al.[93] described this concern by showing that environmental samples tested by molecular methods such as RT-PCR tested positive for inactivated viruses. Discharge of tertiary treated sewage effluent in the upstream sites will also contribute to the PCR detectable but nonviable viral genomes into the surface waters.

In comparing the seasonal distributions of fecal indicator bacteria and coliphages, the highest concentrations of all parameters were detected in November during a storm period. In addition, similar trends were observed with high concentrations in February during a period of heavy rain. No significant change was observed in January, indicating that light rain showers may not have a significant impact on urban runoff. In contrast, the seasonal distribution of adenoviruses showed a different pattern, suggesting the source of fecal indicator bacteria and human viruses are different.

No positive results for adenovirus from the infectivity assay were observed although approximately 16% of samples were positive for adenovirus by real-time PCR, with some sample measuring up to 10^4 genome equivalence per liter of water. It is conceivable that the adenoviruses detected in the environmental samples may not be infectious with any binding affinity to A459 and HEK-293A cells. In addition, the tissue cultures used in this study may not be sensitive for specific serotypes present in the environmental samples. Environmental inhibitors may also be a factor because viruses are known to adsorb to solids such as sediments, which may prevent their detection, thereby resulting in a false negative.

This study presents the first quantitative investigation of human adenoviruses in environmental samples using real-time PCR and represents a significant advancement in the ability for rapid assessment of the viral quality of waters. However, PCR results significantly overestimated the occurrence of infectious viruses in the environment. Therefore, caution should be taken at directly extrapolating positive PCR results of human viruses to human health risks. Furthermore, human adenoviruses were more frequently detected than enteroviruses, suggesting their potential to serve as a better index for human sewage.

CHAPTER 11.0

Comparison of Four Tissue Culture Cell Lines for Detection of Human Adenoviruses in Environmental Samples

11.1 Summary

Human viral contamination of drinking and recreational waters poses an important human health risk. To prevent human exposure, source waters used for drinking water supply and recreational waters must be monitored for viral contamination. In recent years, molecular approaches for viral detection such as PCR have overtaken the traditional tissue culture assay due to the improved sensitivity of detection. However, these molecular methods often yield little or no information on viral infectivity and often are insufficient for regulatory and policy decisions. In comparison with the efforts in molecular technologies, little energy has been applied to testing and improving tissue culture cell lines for detecting human viruses in the environment. Here the researchers compared four different human cell lines, including A549 (human lung carcinoma cells), 293A (human embryonic kidney cells, continuously expressing adenovirus E1 protein, which can enhance adenovirus replication), HepG2, and Huh7 (both human liver carcinoma cells), for their infectivity and sensitivity to human adenovirus serotypes 40 and 2. The sensitivity (as measured by TCID50) of 293A appears to be higher than all other cell lines used, especially to fastidious serotype 40. A549 was the least sensitive to serotype 40 but showed some degree of sensitivity to serotype 2. HepG2 and Huh7 were sensitive to infection by both adenovirus serotypes 2 and 40 but less so than 293A cells. Application of 293A and A549 cell lines to over 200 water samples collected from Newport Bay, Calif., a small estuary surrounded by highly urbanized area, showed that positive samples were only detected using 293A cell line but none was positive by A549 cells. Among 10 sewage samples tested using the two above cell lines, positive results of human viral infection were also exclusively detected in 293A cells.

This result suggests that 293A cell line is the best for environmental assessment of human adenovirus contamination. Therefore, this research is an important contribution to the effort of environmental monitoring of viruses in waters.

11.2 Introduction

Enteric adenoviruses 40 and 41 are important waterborne viruses, and many studies identify them as the second most important agents of gastroenteritis in children next to rotaviruses. Adenoviruses were recently placed on the U.S. EPA drinking water contaminant candidates list. The standard method for detection of viral pathogens in environmental samples uses assays in mammalian cell culture. However, adenoviruses 40 and 41 are difficult to culture and do not produce clear and consistent cytopathogenic effects using the commonly available cell lines (A549). Cell line HEK-293A is a genetically modified continuous line of transformed human embryonic kidney cells. This cell line contains an insertion of gene sequences encoding for adenovirus early expression protein (E1 protein), which consecutively express in HEK-293A. Compared with the traditional 293 cell line, 293A cell line has the phenotype that easily adheres to the surface of a culturing flask and forms mono layers of tissue. Brown et al. (1986) demonstrated 293A's sensitivity in the isolation of adenoviruses, particularly the fastidious species 40 and 41 from stool specimens. However, this genetically modified cell line has not been applied in environmental water research. The goal of this study was to compare the sensitivity of four different tissue culture cell lines for recovery of human adenoviruses from environmental samples.

11.3 Materials and Methods

11.3.1 Cell Lines and Culture Conditions

Human embryonic kidney cells HEK-293A and human liver carcinoma cells HepG2 and Huh 7 were obtained from the University of Southern California (curtsey of Professor Michael Lai). The cells were grown at 37°C in Dulbecco's modified essential medium (DMEM) containing 5% fetal calf serum (FBS), 2 mM L-glutamine, 100 U/ml penicillin and 100 mcg/ml streptomycin. Human lung carcinoma cell line (A549) was obtained from Los Angeles Sanitation District at 108-passage and was used in this study between 115- to 123-passage. A549 cells were grown in Ham's F12 medium, also containing 5% FBS, 2 mM L-glutamine, 100 U/ml penicillin and 100 mcg/ml streptomycin. Cells were subcultured at four- to five-day intervals with a trypsin EDTA solution (Cellgro).

11.3.2 Human Adenoviruses

Adenovirus serotype 40 (ad-40) was isolated from an environmental sample of Newport Bay, CA by cell culture (HEK-293A) in the laboratory and confirmed by sequencing of a 482-bp region in the hexon gene (see Chapter 6.0 for details of isolation and sequencing). Adenovirus serotype 2 was obtained from Los Angeles Sanitation District.

11.3.3 Sensitivity Assay Using TCID50

To compare the sensitivity of different cell lines to adenovirus serotypes 2 and 40 infection, TCID50 test was conducted using each of four cell lines. The experimental procedure is schematically outlined in Figure 11-1. In brief, cells were cultured to confluent, then each 100 μl of serial diluted viral stock was added to the cell surface and incubated for 10 days. CPE was observed under microscopy. The number of positive wells per dilution was counted and used for calculation of TCID50.

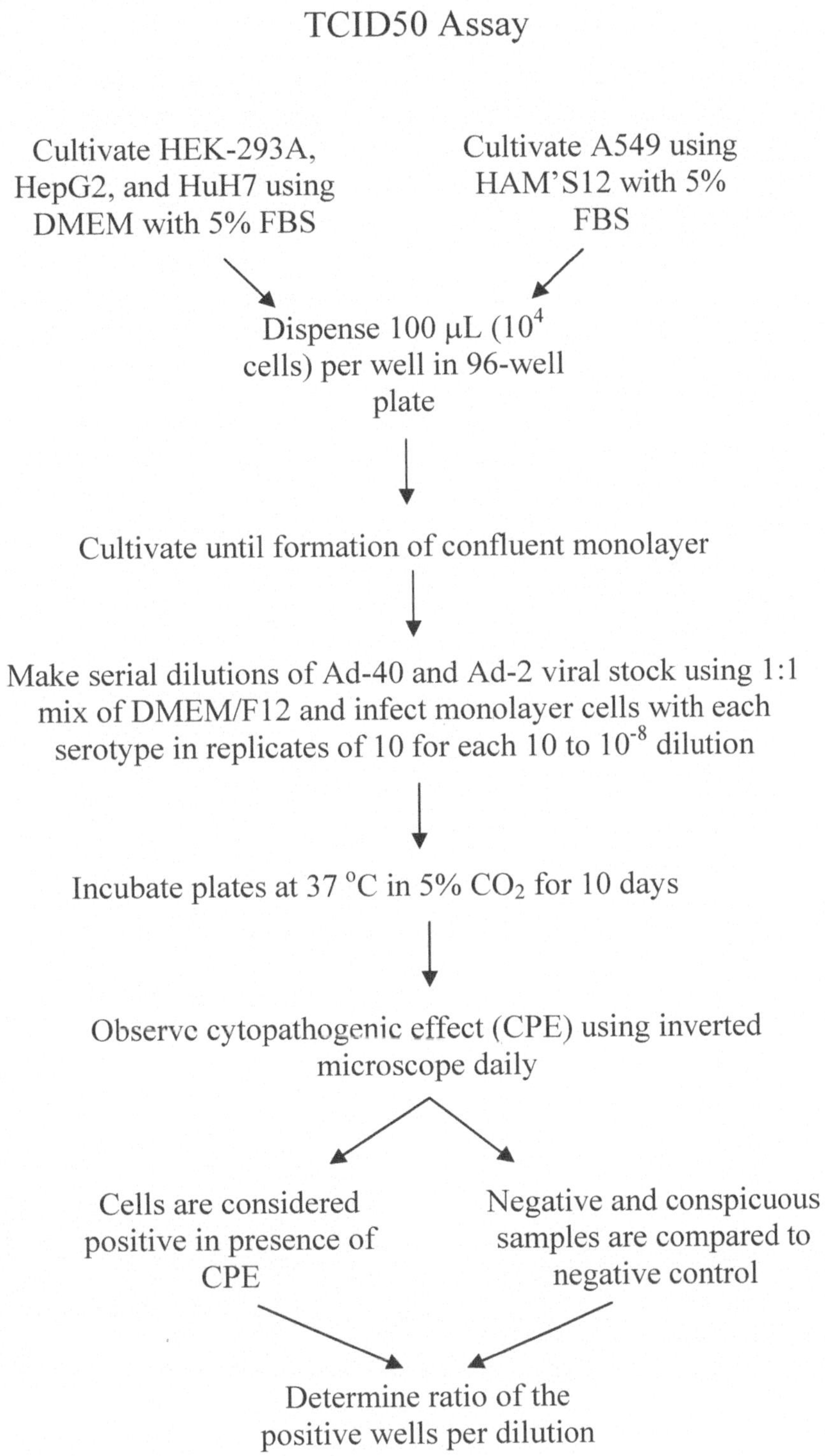

Figure 11-1. Schematic Representation of Method for Comparison of the Sensitivity to Human Adenoviruses Infection Using Four Different Tissue Culture Cell Lines.

11.4 Results

Figure 11-2 shows the results of the sensitivity assay using four different cell lines with two different serotypes of human adenoviruses. The sensitivity (as measured by TCID50) of 293A appears to be higher than all other cell lines tested, especially to fastidious serotype 40. A549 was the least sensitive to serotype 40 but showed similar sensitivity to serotype 2 at the end of the 10-day incubation period. The infectivity of ad-2 was delayed for two days on A549 cells, likely due to the advantage of existing E1 protein produced by the 293A cell line to support the replication of adenoviruses. HepG2 and Huh7 were sensitive to infection by both adenovirus serotypes 2 and 40, but the TCID50 value on HepG2 cells was 4 logs below those found on 293A cells, and on Huh7 was 5 logs below.

Application of 293A and A549 cell lines to over 200 water samples collected from Newport Bay, CA showed that ~40% of samples were positive for human viruses (including ad-40 confirmed by sequencing; see Chapter 6.0 for detailed plaque assay) using 293A cell line. However, none were positive on A549 cells (Figure 11-3). Among 10 sewage samples tested using the two above cell lines, positive results of human viral infection were also exclusively detected in 293A cells.

Adenoviruses serotype 2

Adenoviruses serotype 40

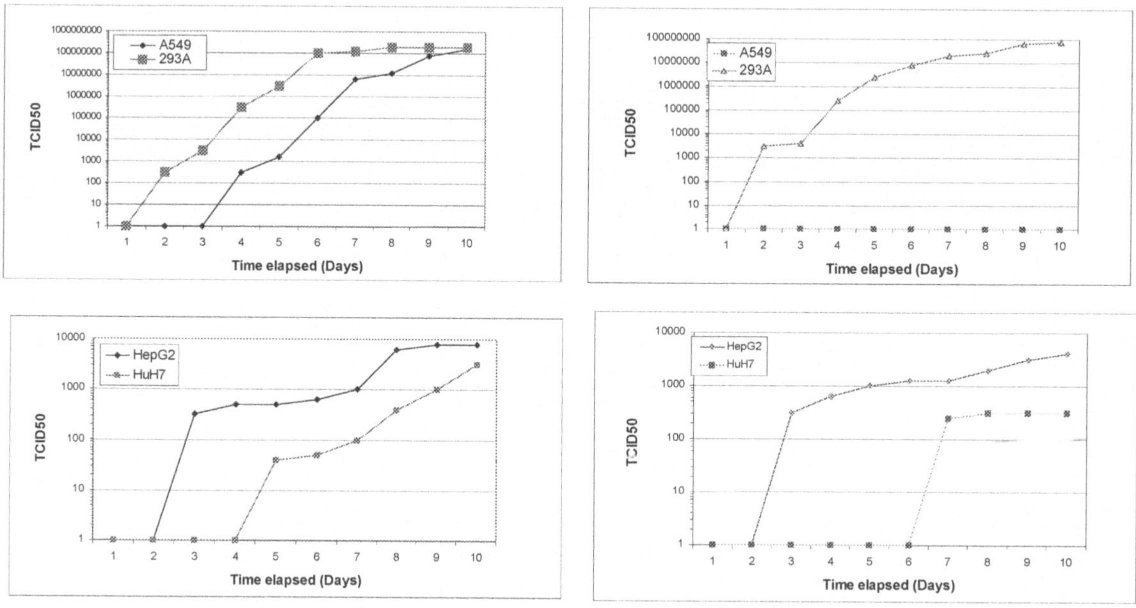

Figure 11-2. Comparison of Four Tissue Culture Cell Lines for The Sensitivity for Human Adenovirus 2 and 40 Infection.

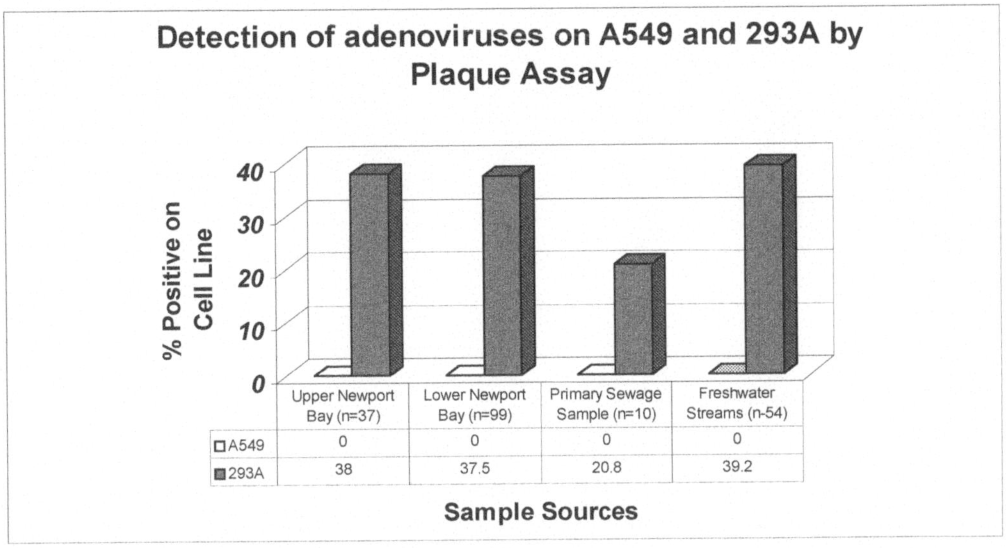

Figure 11-3. Detection of Human Adenoviruses in Environmental Samples on Two Different Tissue Culture Cell Lines

11.5 Discussion and Conclusion

This study demonstrated that human embryonic kidney cell 293A is the most sensitive cell line to adenovirus infection by both serotypes 2 and 40. However, tissue culture cell line A547, which is commonly used for viral quality assay of sewage effluent, is insensitive to infection by adenovirus serotype 40. Infectious human viruses were only recovered using cell line 293A, suggesting that most of the environmental isolates perhaps belong to serotype 40. No infectious virus was isolated on cell line A549 in 200 environmental samples tested including primary sewage. Overall, this study showed that 293A cell line is the best suited for environmental assessment of human adenoviruses contamination. Therefore this research is an important contribution to the effort of environmental monitoring of viruses in water.

CHAPTER 12.0

REAL-TIME PCR QUANTIFICATION OF ADENOVIRUS IN SEDIMENTS

12.1 Introduction

Sediments in the natural system were suspected as a source of water column microbial contamination. Studies have shown that viruses and bacteria can be protected from environmental degradation conditions such as solar radiation when in association with particulates in water[94]. The solid-associated bacteria and viruses also tend to settle onto the bottom sediments where they reach high concentrations and survive for longer periods of time than in the water column[95]. Previous studies have shown that resuspension of the upper layer of sediments by natural and artificial phenomena can lead to an increase of virus concentration in the water column[96, 97]. However, little information is available on the quantitative detection of viruses in aquatic sediments. The effort to develop viral extraction protocols from sediment and real-time PCR quantification is reported here.

12.2 Materials and Methods

12.2.1 Sediment and Sludge Samples

Marine sediments used in this study were fine, sandy sediment collected from Newport Dunes, Newport Bay, Calif. Preliminary investigations of fecal indicator bacteria in these sediments indicated 10 to 100 times greater concentrations than those observed in the water column. Sediments were collected using sediment grab from a small boat and transferred into a 50-ml sterile tube. All sediment samples were transported on ice to UCI lab immediately and stored at -80°C until used. Sewage sludge samples were obtained from the Orange County Sanitation District wastewater treatment facility.

12.2.2 Seeding Viruses in Sediments

To mimic viral contaminated aquatic sediments, the researchers first seeded the sediments or sewage sludge with human adenoviruses. One gram of sediment/sludge was seeded with 1 ml of 10^5/ml recombinant adenovirus serotype 5 (rAd-5) or environmental isolates of adenovirus serotype 40 (Ad40) buffered in PBS. The sediment slurries were then incubated at 4°C with gentle shaking at 400 rpm to allow adsorption of viruses to sediment particles. A time-course study was carried out to investigate the time required for viral adsorption to sediments. Incubation times of 2, 6, and 18 hours were tested. After adsorption, the sediment and viral lysate mixtures were then centrifuged at 3000x g for 5 min. The supernatants were poured out and assayed for free viruses using tissue culture plaque assay, real-time PCR assay, and nested PCR assay after viral nucleic acid extraction. The quantity of unattached viruses was used to determine viruses that adsorbed onto the sediment. The sediment pellets were used for developing protocols for viral extraction from sediments for real-time PCR assay.

12.2.3 Developing Protocols for Viral Extraction from Sediments

To develop optimal protocols for viral extraction from sediments suitable for real-time PCR, the researchers first experimented with previously reported extraction methods for marine sediment and sewage sludge [98-100] with the following modifications. In brief, 5 ml of 10% beef extract (pH 9) was added to the sediment pellets. The mixture was shaken at 400 rpm for 30 min, then sonicated with a microprobe (Sonic Dismembrator model 500 from Fisher Scientific, Sonicator 550 and 3000 from Misonx) on ice for either 2 min or 1 min, with a 20-sec "pulse on" and a 40-sec "pulse off." The samples were then centrifuged at 3000 g for f5 min at 4°C. The supernatants were adjusted to pH 7.2 with HCl, and viruses in the supernatant were precipitated by adding 8% (w/v) polyethylene glycol 8000. The mixture was mixed by gentle shaking at 400 rpm for 2 hours at 4°C, then incubated overnight at 4°C without shaking to allow precipitation of viruses. The next morning samples were centrifuged at 10,000 g for 90 min at 4°C. The pellet was then collected and resuspended in 0.5 ml of 1x phosphate buffer (KH2PO4 0.144g/L, NaCl 9.00g/L, NaHPO4 0.795g/L pH 7.4±0.1). The beef extract-eluted samples were decontaminated by chloroform extraction, using an equal volume of chloroform. The extracts were used for real-time PCR assay (as described in the Chapter 5.0), nested-PCR assay[18], and plaque assay using HEK-293A cell-line as previously described (Chapter 2.0 and Chapter 6.0). The efficiency of viral recovery was calculated by comparing the quantity in the final extracts with seeded viral concentration.

12.2.4 Nucleic Acid Purification to Remove PCR Inhibitors from Sediment Extracts

During the development of the viral extraction protocol, it was observed that the presence of real-time PCR inhibitors in the final extracts was a major obstacle for efficiently detecting viral signal. To further purify the viral nucleic acid extracts from sediment, the researchers experimented with the viral DNA purification kit from Qiagen and the Power Soil purification kit from Bio101 following manufacturer's instructions. In addition, they adapted the protocols of Holben et al.[101] and Berthelet et al.[102] using PVPP (polyvinylpolypyrrolidone) spin columns as previously described. To fulfill the need of small final volume for real-time PCR assay, the researchers modified the original protocols by adding an ethanol precipitate step to concentrate the target DNA.

12.3 Results and Discussion

Time-course study to determine incubation time required for viral seeding in sandy sediments showed that mixing adenovirus with sediments for 2 hours resulted in similar adsorption efficiency as incubating for 6 and 18 hours (Table12-1). Over 99% of viruses were bond to the surface of the sediment. Therefore, all sediments used for viral extraction protocol development were incubated with seeding viral stocks for 2 hours at 4°C before separation of sediments from supernatants by low speed centrifugation. The artificially contaminated sediment pellets were used for the viral extraction protocol development described below.

Table 12-1. Relationship Between Mixing Time and Binding Efficiency of Adenovirus to Sediments. (10^6 adenoviruses were Initially added to the sandy sediment)

Time of Mixing	Copies in the Supernatant	Binding Efficiency
2 Hours	10^4	99%
6 Hours	10^4	99%
Overnight	10^4	99%

Table 12-2 shows the results of adenovirus recovery from sediments using two different sonication time settings. There was no significant difference in these two settings used. In general, real-time PCR was capable of detecting seeded viruses within the same log range of initial seeding concentration in sandy sediments. This result was similar to that detected by traditional nested-PCR quantified by dilution serials and infectivity by plaque assay. The average efficiency of viral recovery from sediment detected by real-time PCR was 11.3%, while the recovery by plaque assay was 7.3%. However, real-time PCR in most cases failed to detect adenoviruses seeded in the sludge samples, while positive results were obtained by traditional nested-PCR and in some cases by plaque assay. The most possible explanation for this result is that real-time PCR is more sensitive to the presence of PCR inhibitors in the reaction.

To remove PCR inhibitors from sludge and marine sediments, the researchers adapted a double washing step using QIAamp viral RNA Mini Kit for the adenoviruses DNA purification. According to the manufacturer's instruction, a second washing step may improve the purity of the final nucleic acid and removal of PCR inhibitors. Figure 12-1 shows the result of double washing using QIAamp Viral Kit for seeded sandy sediment samples. The average recovery detected by real-time PCR increased slightly from 8.8% to 10%. However, this modified protocol still produced negative results for seeded sludge samples by real-time PCR. To further purify DNA, the researchers adapted the protocol of Holben et al.[101] and Berthelet et al.[102] with PVPP (polyvinylpolypyrrolidone) spin columns. They also tested another commercial soil extraction kit, PowerSoil DNA Kit (Bio101), for adenovirus DNA purification from sludge samples. Only the soil kit extracted DNA-yield positive results for sludge samples using real-time PCR method (Figure 12-2).

Table 12-2. Recovery of Adenoviruses from Sediments and Sludge Detected by Real-time PCR, Traditional PCR, and Plaque Assay.

Experimental Trial	Matrix	Sonication Period	% Rec. by Real-Time PCR	Detection by Traditional PCR	% Rec. by Plaque Assay
1	Sandy Sediments	2 min	15.4	+	0.2
2	Sandy Sediments	2 min	4.8	+	31.8
3	Sandy Sediments	2 min	18.4	+	6.7
4	Sandy Sediments	2 min	10.1	+	0.3
5	Sandy Sediments	1 min	25.4	+	12.9
6	Sandy Sediments	1 min	4.8	+	0.0
7	Sandy Sediments	1 min	4.1	+	4.9
8	Sandy Sediments	1 min	7.9	+	1.6
Average			11.3		7.3
Standard Deviation			7.7		10.8
9	Sludge	2 min	0.9	+	Tox
10	Sludge	2 min	0.0	+	0.2
11	Sludge	2 min	0.0	+	1.5

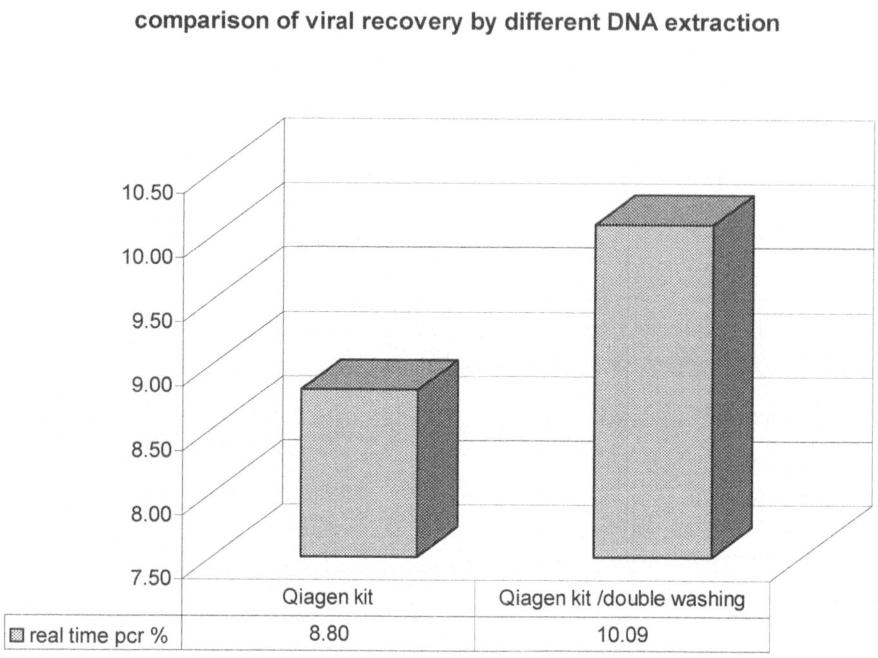

Figure 12-1. Comparison of Adenovirus Recovery from Sandy Sediments Detected by Real-time PCR Using QIAamp Viral Kit with Double Washing Steps.

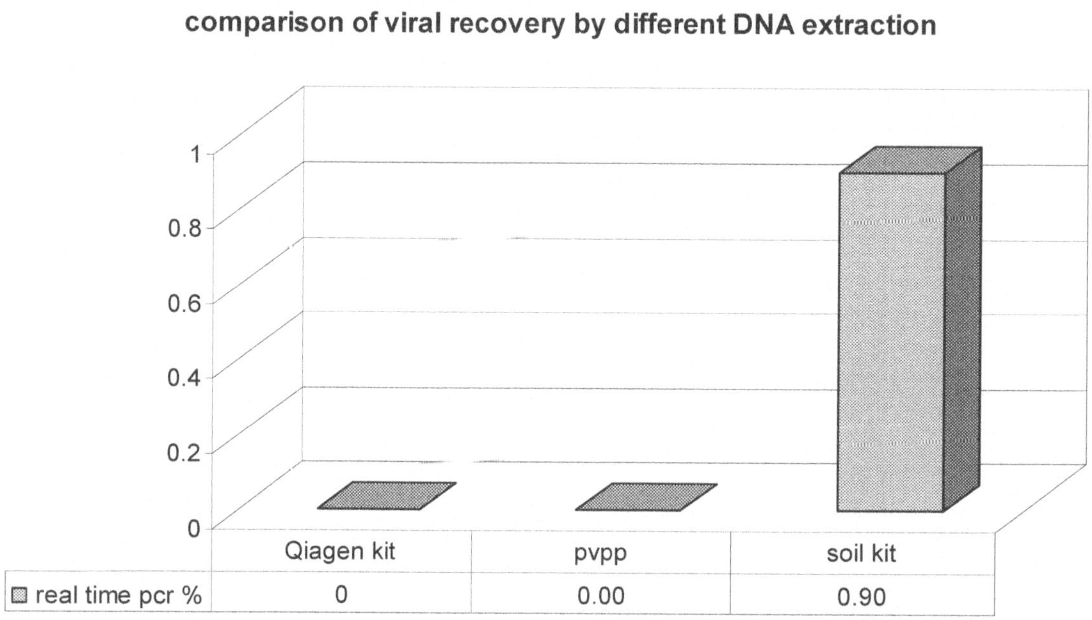

Figure 12-2. Comparison of Effects of Different Viral DNA Extraction Methods for Recovery of Seeded Adenoviruses from Sludge Detectable by Real-time PCR.

12.3 Conclusions

In this study, an adenovirus (rAd-5) extraction protocol was developed to determine the rate of recovery of adenoviruses from aquatic sediment and sewage sludge samples. Preliminary research suggests that the real-time PCR based method for viral detection can be applied to aquatic sediment samples containing adenoviruses. However, this method may not be suitable for sewage sludge samples due to the presence of high concentration of PCR inhibitors. However, this problem is not unique to this study; detection of viral load in sediment and sludge samples continues to present challenges to researchers worldwide. Further improvement of viral DNA extraction and purification methods will contribution significantly to the ability to quantify viral contamination in sediments.

CHAPTER 13.0

SUMMARY AND CONCLUSIONS: MOLECULAR DETECTION OF PATHOGENS AND INDICATORS—ADVANTAGES AND PITFALLS

The advancement of molecular biology and its application in environmental microbiology has represented significant advancement in the ability to rapidly detect microbial pathogens in the environment. Reviewing the history of pathogen detection, molecular biology based alternative methods have improved the speed of detection from multiple days down to a few hours. Using human adenoviruses as an example, the real-time PCR-based method developed in this study has allowed the rapid amplification of target genome within 4 to 6 hours, while the tissue culture method requires seven to 14 days. The real-time PCR method developed and reported in this study not only allows for rapid and specific detection of human adenoviruses but also yields quantitative estimation of viral concentration in the environment.

Furthermore, molecular techniques provide a means for detecting pathogens that are hard to culture. The PCR method is the only method currently available to specifically identify norovirus because tissue culture cell lines to support the replication of this virus have not been discovered. In this report, quantitative PCR detection of enterococci also sheds light on the current debate over the occurrence of noncultivable organisms in the environment. In the past 20 years, rcsearchers have come a long way from culture-based pathogen and indicator detection to the full realization of the potential of molecular alternatives. On the other hand, the development of new molecular alternatives also presents additional questions and challenges that can not be answered by molecular approaches alone. For example, the real-time PCR detection of *Enterococcus* in chlorinated sewage samples yielded similar results to those untreated by chlorine, while the culture method indicated that chlorine sufficiently inactivated the infectious *Enterococcus*. This discrepancy indicates that microbial detection that is based solely on detecting a small piece of genome material is not a good indication of the infectivity of the target organism.

In theory, incorporating membrane or capsid integrity and genome copies quantified by real-time PCR may reduce the possibility of detecting fragments of genomic material. In this study, however, attempts to use a magnetic bead anti-adenovirus-antigen complex to capture physically intact adenovirus particles for downstream PCR detection were not successful. Further investigations of the physical, chemical, and biological properties of adenoviruses may resolve the problems encountered during this study.

An alternative way to rapidly assay the infectivity of viral pathogens in the environment is to combine the traditional tissue culture method with the molecular biology approach. Integrated tissue culture-PCR has proven successful for detecting infectious human viral pathogens in several reports. In this study, a new cell line for isolation of environmental adenoviruses was investigated. Cell line 293A, traditionally used in clinical research for adenovirus packaging, has proven efficient for isolating viruses from the environment. Combining this tissue culture cell line with reverse transcriptase PCR amplification will contribute to the ability to rapidly detect adenovirus in environmental samples.

The question and answer format below aims to provide a rounded view of the advantages and pitfalls of molecular alternatives for pathogen and indicator detection. The questions below are raised with the intention of communicating the outcome of this study to a broader audience regarding the advantages and shortcomings of molecular alternatives for human viruses and pathogen detection. The researchers believe these questions provide an excellent summary of what has been learned from this project.

Q: Why is it important to switch to the direct measurement of viral concentration for environmental monitoring?

A: Viruses behave very differently from fecal indicator bacteria during the sewage treatment process. They are not sufficiently eliminated using the procedures that are sufficient to remove fecal bacteria. Once released into the environment, viruses are also more resistant than fecal bacteria to degradation conditions such as salt concentration in seawater or solar radiation. Direct measurement of viral concentration will better represent the microbial quality of discharging wastewater and receiving water. Viral concentration in receiving water will also provide references to the health risks people run when exposed to contaminated waters.

Q: Why is it taking the environmental community so long to switch from indicators to direct measurement of pathogens?

A: Technology difficulties are one of the major hurdles to overcome. The current methods for direct pathogen measurement are still at the research stage. The direct transformation of these technologies requires further development of streamlined procedures. For example, the different levels of inhibitor presented in the environment samples may prevent the positive detection of the viral target. The variability of the environmental matrices also demands protocols be developed to fit individual needs. Furthermore, new regulations and policies should also be implemented before the switch can truly take place.

Q: Will indicators eventually be discarded for direct pathogen measurement?

A: No. Measuring all three indicator bacteria, namely total coliform, fecal coliform, and *Enterococcus* (currently used by many states in the U.S.) may not be necessary. However, indicators provide the basic guideline for in-depth background knowledge and decades of data to compare with.

Q: How much quicker is real-time PCR than fecal coliform analysis?

A: Real-time PCR from nucleic acid extraction to results reporting takes about four hours, while fecal coliform analysis takes about a day and half (30 hours).

Q: What is the cost difference? Will that difference shrink over time?

A: Currently, real-time PCR costs about ten times more than traditional culture assay of fecal coliform. The researchers do believe the difference will shrink over time.

Q: Where would this technology be the most applicable? At WWTPs? Beach testing?

A: The technology is applicable to both areas mentioned as well as to water used as a source for drinking water.

Q: How field ready is this technology?

A: It is not field ready at the current stage. However, it can be field ready within the next five years. Additional research should be focused on understanding the relationship between real-time PCR positive results and viral infectivity and further improving target concentration and purification technology to deal with complex environmental matrices. More importantly, the researchers think the major problem is not how to have this method field ready but rather to have the policy guidelines to follow up with the new method. It is very hard for end users to switch to a new testing method without policy guidance. Most end users are already overworked and understaffed, so it will be difficult for them to pick up another challenging test without any guidelines.

Q: The report indicates that the tests designed only detect human viruses. Are those the only ones that can make people ill?

A: The viruses targeted in this study are human pathogens. However, the population susceptibility to these viruses is very different. Children are most susceptible; adults exposed to these viruses may not display any disease symptoms. The virus used in this study is meant to be used as an index for the occurrence of other pathogenic viruses, but they are not the only ones that make people ill. There are more than a hundred others that can also make people ill.

Q: The report says that genome-based detection technology overestimates the concentration of infectious viruses in the environment. Why is that so?

A: The genome-based technology (PCR) only looks at the integrity of a piece of nucleic acid. Viruses with structural damage may be also detected by this method, although the damaged viruses can not replicate in the human body.

Q: Does the report include the directions to perform this method?

A: Yes, the report gives all necessary protocols to perform this method in other labs.

Q: How reliable is this method at this point?

A: If the QA/QC procedures are adequate, the method will not produce false positive results. However, the method may produce false negative results if the target concentration is below the lower detection limit or the presence of PCR inhibition in the assay reduces the efficiency of amplification.

Q: How does one interpret quantitative PCR results that indicate the presence of human viruses in a water sample?

A: Quantitative PCR positive results of human viruses only indicate the presence of human viral genome and yield little information regarding the infectivity of these viruses. The information should be used as one of the multiple tools in a tool box to assess the quality of water.

Q: What do the PCR positive results mean in the context of human health risk?

A: The positive PCR results by themselves do not indicate human health risk. However, the results reflect the presence of human waste contamination in the water. The human health risk should be assessed using multiple lines of evidence. For example, if the presence of human viral signal is accompanied by higher than normal concentrations of traditional indicators based on culture assay, it may indicate a sewage spill or discharge of improperly treated sewage effluents. Under this circumstance, the risk of human exposure to pathogens is high.

ENDNOTES

[1] WHO "Guidelines for drinking water quality," World Health Organization, 2004.
[2] Frahm, E.; Obst, U.J Microbiol Methods 2003, 52, 123-131.
[3] U.S. EPA "Ambient water quality criteria for bacteria-1986.," U.S. Environmental Protection Agency, 1986.
[4] U.S. EPA "National primary drinking water standards," U.S. Environmental Protection Agency, 1994.
[5] Jiang, S.; Noble, R.; Chu, W. *Applied and Environmental Microbiology 2001*, 67, 179-184.
[6] CDC "Water-related disease outbreaks, 1991-1992. Morbid. Mortal. Weekly Report," Centers for Disease Control, 1993.
[7] Berg, G. *Viral pollution of the environment*; CRC Press: Boca Raton, FL, 1983.
[8] DeLeon, R.; Jaykus, L. In *Manual of Environmental Microbiology*; Hurst, C., Knudsen, G. R., McInerney, M.J., Stetzenbach, L.D., Walter., M.V., Eds.; American Society for Microbiology: Washington, D.C., 1997.
[9] Melnick, J.L.; Gerba, C.P. *Crit. Rev. Environ. Control* 1989, 10, 65.
[10] Blacklow, N.; Greenberg, H.B. *New England J. Med.* 1991, 325, 252-264.
[11] Uhnoo, I.; Wadell, G.; Svensson, L.; Olding-Stenkvist, E.; Ekwall, E.; Molby, R. J *Infect* 1986, 13, 73-89.
[12] Cruz, J.R.; Caceres, P.; Cano, F.; Flores, J.; Bartlett, A.; Torun, B. *J Clin Microbiol* 1990, 28, 1780-1784.
[13] Crabtree, K.D.; Gerba, C.P.; Rose, J.B.; Haas, C.N. *Water Science and Technology* 1997, 35, 1-6.
[14] Moe, C. In *Manual of environmental microbiology*; Hurst, C., Knudsen, G., McInerney, M., Stetzenbach, L., Walter, M., Eds.; American Society for Microbiology: Washington, D.C., 1997; 136-152.
[15] Castignolles, N.; Petit, F.; Mendel, I.; Simon, L.; Cattolico, L.; Buffet-Janvresse, C. *Molecular and Cellular Probes* 1998, 12, 175-180.
[16] Chapron, C.D.; Ballester, N.A.; Fontaine, J.H.; Frades, C.N.; Margolin, A.B. Abstracts of the General Meeting of the American Society for Microbiology 2000, 100, 547.
[17] Tani, N.; Dohi, Y.; Kurumatani, N.; Yonemasu, K. *Microbiology and Immunology* 1995, 39, 577-580.
[18] Pina, S.; Puig, M.; Lucena, F.; Jofre, J.; Girones, R. *Applied and Environmental Microbiology* 1998, 64, 3376-3382.
[19] Puig, M.; Jofre, J.; Lucena, F.; Allard, A.; Wadell, G.; Girones, R. *Applied and Environmental Microbiology* 1994, 60, 2963-2970.
[20] Allard, A.; Albinsson, B.; Wadell, G. J. *Med. Virol.* 1990, 37, 149-157.
[21] Irving, L.G., Smith, F.A. *Appl. Environ. Microbiol.* 1981.
[22] Krikelis, V.; Spyrou, N.; Markoulatos, P.; Serie, C. *Can J Microbiol* 1985, 31, 24-25.
[23] Krikelis, V.; Markoulatos, P.; Spyrou, N.; Serie, C. *Wat. Sci. Technol.* 1985, 17, 159-164.
[24] Enriquez, C.E.; Hurst, C.J.; Gerba, C.P. *Water Research* 1995, 29, 2548-2553.
[25] Rao, V.C.; Melnick, J.L. *Aspects of microbiology*; American Society for Microbiology: Washington, D.C., 1986; Vol. 13.

[26] Autho In Water Quality Conference; American Water Works Association, Denver, CO: San Diego, CA, 1990; Vol. 18, 833-853.
[27] Karaganis, J.V.; Larking, E.P.; Melnick, J.L.; Scarpino, P.V.; Schaub, S.A.; Sorber, C.A.; Sullivan, R.; Wellings, F. M. "Research priorities for monitoring viruses in the environment.," U.S. environmental Protection Agency, Office of Research and Development, Environmental Monitoring and Support Laboratory, 1983.
[28] Cabelli, V.J. "Health effect critia for marine waters.," U.S. Environmental Protection Agency, 1983.
[29] Dufour, A.P. "Health effects criteria for freshwater recreational waters.," U.S. Environmental Protection Agency, 1984.
[30] Cabelli, V.J. "Health criteria for marine recreational waters.," Environmental Protection Agency., 1980.
[31] Miescier, J.J.; Cabelli, V.J.J. *Water Pollut. Control Fed.* 1982, 54, 1599-1606.
[32] Regenmortel, M.H.V. v.; Fauquet, C.M.; Bishop, D.H.L.; Carstens, E.B.; Estes, M.K.; Lemon, S.M.; Maniloff, J.; Mayo, M.A.; McGeoch, D.J.; Pringle, C.R.; Wickner, R.B. *Virus Taxonomy: The Classification and Nomenclature of Viruses*. The Seventh Report of the International Committee on Taxonomy of Viruses; Academic press: San Diego, CA. 2000.
[33] Hierholzer, J.C. In *Diagnostic procedures for viral, rickettsial and chlamydial infections*; E. H. Lennette, A.D.L., and T.E. Lennette, Ed.; American Public Health Association, Inc.: Washington, D.C., 1995; 169-188.
[34] Benkö, M.; Harrach, B.; and Russell, W.C. In *Virus taxonomy*. Seventh report of the International Committee on Taxonomy of Viruses.; M.H.V. Van Regenmortel, C.M.F.; Bishop, D.K.L.;Carstens, E.B.; Estes, M.K.; Lemon, S.M.; Maniloff, J.; Mayo, M.A.; McGeoch, D.J.; Pringle. C.R., and Wickner, R.B. Ed.; Academic Press, Inc.,: New York, N.Y., 1999; 227-238.
[35] De Jong, J.C.; Wermenbol, A.G.; Verweij-Uijterwaal, M.W.; Slaterus, K.W.; Wertheim-Van Dillen, P.; Van Doornum, G.J.; Khoo, S.H.; Hierholzer, J.C. *J Clin Microbiol* 1999, 37, 3940-3945.
[36] Gerba, C.P.; Gramos, D.M.; Nwachuku, N. *Appl Environ Microbiol* 2002, 68, 5167-5169.
[37] Boom, R.; Sol, C.J.; Salimans, M.M.; Jansen, C.L.; Wertheim-van Dillen, P.M.; van der Noordaa, J. J Clin Microbiol 1990, 28, 495-503.
[38] Ririe, K.M.; Rasmussen, R.P.; Wittwer, C.T. *Anal Biochem* 1997, 245, 154-160.
[39] Espy, M.J.; Ross, T.K.; Teo, R.; Svien, K.A.; Wold, A.D.; Uhl, J.R.; Smith, T.F., *J Clin Microbiol* 2000, 38, 3116-3118.
[40] Schaade, L.; Kockelkorn, P.; Ritter, K.; Kleines, M. *J Clin Microbiol* 2000, 38, 4006-4009.
[41] Ballard, A. L.; Fry, N. K.; Chan, L.; Surman, S.B.; Lee, J. V.; Harrison, T.G.; Towner, K.J. *J Clin Microbiol* 2000, 38, 4215-4218.
[42] Bellin, T.; Pulz, M.; Matussek, A.; Hempen, H.G.; Gunzer, F.*J Clin Microbiol* 2001, 39, 370-374.
[43] Reischl, U.; Linde, H.J.; Metz, M.; Leppmeier, B.; Lehn, N. *J Clin Microbiol* 2000, 38, 2429-2433.
[44] Taylor, M.J.; Hughes, M.S.; Skuce, R.A.; Neill, S.D. *J Clin Microbiol* 2001, 39, 1272-1278.
[45] Matsumura, M.; Hikiba, Y.; Ogura, K.; Togo, G.; Tsukuda, I.; Ushikawa, K.; Shiratori, Y.; Omata, M. *J Clin Microbiol* 2001, 39, 691-695.
[46] Cane, P.A.; Cook, P.; Ratcliffe, D.; Mutimer, D.; Pillay, D. *Antimicrob Agents Chemother* 1999, 43, 1600-1608.
[47] Torres, M.J.; Criado, A.; Palomares, J.C.; Aznar, J. *J Clin Microbiol* 2000, 38, 3194-3199.
[48] Bustin, S.A. *J Mol Endocrinol* 2000, 25, 169-193.

[49] Thompson, J.D., Gibson, T.J., Plewniak, F., Jeanmougin, F., and Higins, D.G. *Nucleic Acids Research* 1997, 24, 4876-4882.
[50] Saitou, N.; Nei, M. *Mol. Biol. Evol.* 1987, 4, 406-425.
[51] Melnick, J.L. *Prog Med Virol* 1982, 28, 208-221.
[52] Emerson, S.U.; Tsarev, S.A.; Purcell, R.H. *J Hepatol* 1991, 13 Suppl 4, S144-145.
[53] Nainan, O.V.; Margolis, H.S.; Robertson, B.H.; Balayan, M.; Brinton, M.A. *J Gen Virol* 1991, 72 (Pt 7), 1685-1689.
[54] Ross, B.C.; Anderson, D.A. *J Virol Methods* 1991, 32, 213-220.
[55] Tsai, Y.L.; Sobsey, M.D.; Sangermano, L.R.; Palmer, C. J. *Appl Environ Microbiol* 1993, 59, 3488-3491.
[56] De Serres, G.; Cromeans, T.L.; Levesque, B.; Brassard, N.; Barthe, C.; Dionne, M.; Prud'homme, H.; Paradis, D.; Shapiro, C.N.; Nainan, O.V.; Margolis, H.S. *J Infect Dis* 1999, 179, 37-43.
[57] Abd El Galil, K.H.; El Sokkary, M.A.; Kheira, S.M.; Salazar, A.M.; Yates, M.V.; Chen, W.; Mulchandani, A. *Appl Environ Microbiol* 2004, 70, 4371-4374.
[58] Fields, B.N.; Knipe, D.M.; Howley, P.M.; Griffin, D.E. *Fields' Virology*; 4 th ed.; Lippincott Williams & Wilkins: Philadelphia, 2001.
[59] Xu, W.; McDonough, M.C.; Erdman, D.D. *J Clin Microbiol* 2000, 38, 4114-4120.
[60] Tsai, Y.L.; Parker, S.L. *Can J Microbiol* 1998, 44, 35-41.
[61] Emori, T.G.; Gaynes, R.P. *Clin Microbiol Rev* 1993, 6, 428-442.
[62] Moellering, R.C., Jr. *Clin Infect Dis* 1992, 14, 1173-1176.
[63] Patel, R.; Piper, K.E.; Rouse, M.S.; Steckelberg, J.M.; Uhl, J.R.; Kohner, P.; Hopkins, M. K.; Cockerill, F.R., 3rd; Kline, B.C. *J Clin Microbiol* 1998, 36, 3399-3407.
[64] Huycke, M.M.; Sahm, D.F.; Gilmore, M.S. *Emerg Infect Dis* 1998, 4, 239-249.
[65] Devriese, L.A.; Pot, B.; Collins, M.D. *J Appl Bacteriol* 1993, 75, 399-408.
66 Lleo Mdel, M.; Tafi, M.C.; Canepari, P. *Syst Appl Microbiol* 1998, 21, 333-339.
67 Molander, A.; Lundquist, P.; Papapanou, P.N.; Dahlen, G.; Reit, C. *Int Endod J* 2002, 35, 1-6.
[68] Kalinina, O.; Lebedeva, I.; Brown, J.; Silver, *J. Nucleic Acids Res* 1997, 25, 1999-2004.
[69] Santo Domingo, J.W.; Siefring, S.C.; Haugland, R.A. *Biotechnol Lett* 2003, 25, 261-265.
[70] Ibekwe, A.M.; Grieve, C.M. *J Appl Microbiol* 2003, 94, 421-431.
[71] Harms, G.; Layton, A.C.; Dionisi, H.M.; Gregory, I.R.; Garrett, V.M.; Hawkins, S.A.; Robinson, K.G.; Sayler, G.S. *Environ Sci Technol* 2003, 37, 343-351.
[72] Colwell, R.R. and Huq, A. In Vibrio cholerae and cholera: molecular to global perspectives; Wachsmuth, K., Blake, P.A.; and Olsvik, O. Ed.; American Society for Microbiology: Washington, D.C., 1994.
[73] Oliver, J.D.; Bockian, R. *Appl Environ Microbiol* 1995, 61, 2620-2623.
[74] Noble, R.T.; Weisberg, S.B.; Leecaster, M.K.; McGee, C.D.; Dorsey, J.H.; Vainik, P.M.; Orozco-Borbon, V. *Journal of Water and Health* 2003, 1, 23-31.
[75] Cheung, W.H.; Chang, K.C.; Hung, R.P.; Kleevens, J.W. *Epidemiol Infect* 1990, 105, 139-162.
[76] Alexander, L.M.; Heaven, A.; Tennant, A.; Morris, R. *J Epidemiol Community Health* 1992, 46, 340-344.
[77] Toranzos, G.A.; Hanssen, H.; Gerba, C.P. *Water Science and Technology* 1986, 18, 109-114.
[78] Payment, P.; Franco, E. *Applied and Environmental Microbiology* 1993, 59, 2418-2424.
[79] Harm, W. Biological effects of ultraviolet radiation.; Cambridge University Press: Cambridge, United Kingdom, 1980.

[80] Kallenbach, N.R.; Cornelius, P.A.; Negus, D.; Montgomerie, D.; Englander, S. Inactivation of viruses by ultraviolet light.; Basel, Switzerland: Karger, 1989; Vol. 56.
[81] Thurston-Enriquez, J.A.; Haas, C.N.; Jacangelo, J.; Riley, K.; Gerba, C.P. *Applied and Environmental Microbiology* 2003, 69, 577-582.
82 Thurston-Enriquez, J.A.; Haas, C.N.; Jacangelo, J.; Gerba, C.P. *Applied and Environmental Microbiology* 2003, 69, 3979-3985.
[83] Jiang, S.C.; Kellogg, C.A.; Paul, J.H. *Applied and Environmental Microbiology* 1998, 64, 535-542.
[84] Kellogg, C.A.; Rose, J.B.; Jiang, S.C.; Thurmond, J.M.; Paul, J.H. *Mar Ecol-Prog Ser* 1995, 120, 89-98.
[85] Boom, R.; Sol, C.; Beld, M.; Weel, J.; Goudsmit, J.; Wertheim-Van Dillen, P. *Journal of Clinical Microbiology* 1999, 37, 615-619.
86 Giard, D.J.S.; Aaronson, A.; Todaro, G.J.; Arnstein, P.; Kersey, J.H.; Dosik, H.; Parks, W.P. *J. Natl. Cancer Inst*. 1973, 51.
[87] Smith, C.D.; Craft, D.W.; Shiromoto, R.S.; Yan, P.O. *Journal of Clinical Microbiology* 1986, 24, 265-268.
[88] Graham, F.L.; Smiley, J.; Russell, W.C.; Nairn., R.J. *Gen. Virol.* 1977, 36, 59-74.
[89] Goldsmith, K.T.; Dion, L.D.; Curiel, D.T.; Garver, R.I., Jr. *Virology* 1998, 248, 406-419.
[90] Noble, R.T.; Weisberg, S.B.; Leecaster, M.K.; McGee, C.D.; Ritter, K.; Walker, K.O.; Vainik, P.M. *Environ Monit Assess* 2003, 81, 301-312.
[91] Lipp, E.K.; Kurz, R.; Vincent, R.; Rodriguez-Palacios, C.; Farrah, S.R.; Rose, J.B. *Estuaries* 2001, 24, 266-276.
[92] Jiang, S.C.; Chu, W. *J Appl Microbiol* 2004, 97, 17-28.
[93] Sobsey, M.D.; Battigelli, D.A.; Shin, G.A.; Newland, S. *Water Science and Technology* 1998, 38, 91-94.
[94] Landry, E.F.; Vaughn, J.M.; Vicale, T.J.; Mann, R. *Applied and Environmental Microbiology* 1983, 45, 238-247.
[95] Labelle, R.L.; Gerba, C.P. *Applied and Environmental Microbiology* 1979, 38, 93-101.
[96] Bitton, G.; Chou, Y.J.; Farrah, S.R. *Journal of Virological Methods* 1982, 4, 1-8.
[97] Monpoeho, S.; Maul, A.; Mignotte-Cadiergues, B.; Schwartzbrod, L.; Billaudel, S.; Ferre, V. *Appl Environ Microbiol* 2001, 67, 2484-2488.
[98] Middelboe, M.; Glud, R.N.; Finster, K. *Limnol Oceanogr* 2003, 48, 1447-1456.
[99] Chung, H.; Sobsey, M.D. *Water Science and Technology* 1993, 27, 425-428.
[100] Lewis, G.D.; Metcalf, T.G. *Applied and Environmental Microbiology* 1988, 54, 1983-1988.
[101] Holben, W.E.; Jansson, J.K.; Chelm, B.K.; Tiedje, J.M. *Applied and Environmental Microbiology* 1988, 54, 703-711.
[102] Berthelet, M.; Whyte, L.G.; Greer, C.W. *Fems Microbiol Lett* 1996, 138, 17-22.

WASTEWATER UTILITY

Alabama
Montgomery Water Works & Sanitary Sewer Board

Alaska
Anchorage Water & Wastewater Utility

Arizona
Gila Resources
Glendale, City of, Utilities Department
Mesa, City of
Peoria, City of
Phoenix Water Services Department
Pima County Wastewater Management

Arkansas
Little Rock Wastewater Utility

California
Calaveras County Water District
Central Contra Costa Sanitary District
Corona, City of
Crestline Sanitation District
Delta Diablo Sanitation District
Dublin San Ramon Services District
East Bay Dischargers Authority
East Bay Municipal Utility District
Eastern Municipal Water District
El Dorado Irrigation District
Fairfield-Suisun Sewer District
Fresno Department of Public Utilities
Inland Empire Utilities Agency
Irvine Ranch Water District
Las Virgenes Municipal Water District
Livermore, City of
Lodi, City of
Los Angeles, City of
Napa Sanitation District
Orange County Sanitation District
Palo Alto, City of
Riverside, City of
Sacramento Regional County Sanitation District
San Diego Metropolitan Wastewater Department, City of
San Francisco, City & County of
Sanitation Districts of Los Angeles County
San Jose, City of
Santa Barbara, City of
Santa Cruz, City of
Santa Rosa, City of
South Bayside System Authority
South Coast Water District
South Orange County Wastewater Authority
Stege Sanitary District
Sunnyvale, City of
Union Sanitary District
West Valley Sanitation District

Colorado
Aurora, City of
Boulder, City of
Colorado Springs Utilities
Greeley, City of
Littleton/Englewood Water Pollution Control Plant
Metro Wastewater Reclamation District, Denver

Connecticut
The Mattabassett District
New Haven, City of, WPCA

District of Columbia
District of Columbia Water & Sewer Authority

Florida
Broward, County of
Fort Lauderdale, City of
JEA
Miami-Dade Water & Sewer Authority
Orange County Utilities Department
Reedy Creek Improvement District
Seminole County Environmental Services
St. Petersburg, City of
Stuart Public Utilities
Tallahassee, City of
Tampa, City of
Toho Water Authority
West Palm Beach, City of

Georgia
Atlanta Department of Watershed Management
Augusta, City of
Clayton County Water Authority
Cobb County Water System
Columbus Water Works
Fulton County
Gwinnett County Department of Public Utilities
Savannah, City of

Hawaii
Honolulu, City & County of

Idaho
Boise, City of

Illinois
American Bottoms Wastewater Treatment Plant
Greater Peoria Sanitary District
Kankakee River Metropolitan Agency
Metropolitan Water Reclamation District of Greater Chicago
Wheaton Sanitary District

Iowa
Ames, City of
Cedar Rapids Wastewater Facility
Des Moines, City of
Iowa City

Kansas
Johnson County Unified Wastewater Districts
Unified Government of Wyandotte County/ Kansas City, City of

Kentucky
Louisville & Jefferson County Metropolitan Sewer District

Louisiana
Sewerage & Water Board of New Orleans

Maine
Bangor, City of
Portland Water District

Maryland
Anne Arundel County Bureau of Utility Operations
Hagerstown, City of
Howard County Department of Public Works
Washington Suburban Sanitary Commission

Massachusetts
Boston Water & Sewer Commission
Upper Blackstone Water Pollution Abatement District

Michigan
Ann Arbor, City of
Detroit, City of
Holland Board of Public Works
Lansing, City of
Saginaw, City of
Wayne County Department of Environment
Wyoming, City of

Minnesota
Rochester, City of
Western Lake Superior Sanitary District

Missouri
Independence, City of
Kansas City Missouri Water Services Department
Little Blue Valley Sewer District
Metropolitan St. Louis Sewer District

Nebraska
Lincoln Wastewater System

Nevada
Henderson, City of
Reno, City of

New Jersey
Bergen County Utilities Authority
Ocean, County of
Passaic Valley Sewerage Commissioners

New York
New York City Department of Environmental Protection

North Carolina
Charlotte/Mecklenburg Utilities
Durham, City of
Metropolitan Sewerage District of Buncombe County
Orange Water & Sewer Authority

Ohio
Akron, City of
Butler County Department of Environmental Services
Columbus, City of
Metropolitan Sewer District of Greater Cincinnati
Northeast Ohio Regional Sewer District
Summit, County of

Oklahoma
Tulsa, City of

Oregon
Clean Water Services
Eugene/Springfield Water Pollution Control
Water Environment Services

Pennsylvania
Philadelphia, City of
University Area Joint Authority

South Carolina
Charleston Commissioners of Public Works
Mount Pleasant Waterworks & Sewer Commission
Spartanburg Sanitary Sewer District

Tennessee
Cleveland, City of
Knoxville Utilities Board
Murfreesboro Water & Sewer Department
Nashville Metro Water Services

Texas
Austin, City of
Dallas Water Utilities
Denton, City of
El Paso Water Utilities
Fort Worth, City of
Gulf Coast Waste Disposal Authority
Houston, City of
San Antonio Water System
Trinity River Authority

Utah
Salt Lake City Corporation

Virginia
Alexandria Sanitation Authority

WERF SUBSCRIBERS

Arlington, County of
Fairfax County
Hampton Roads Sanitation District
Henrico, County of
Hopewell Regional Wastewater Treatment Facility
Loudoun County Sanitation Authority
Lynchburg Regional WWTP
Prince William County Service Authority
Richmond, City of
Rivanna Water & Sewer Authority

Washington
Everett, City of
King County Department of Natural Resources
Seattle Public Utilities
Sunnyside, Port of
Yakima, City of

Wisconsin
Green Bay Metro Sewerage District
Kenosha Water Utility
Madison Metropolitan Sewerage District
Milwaukee Metropolitan Sewerage District
Racine, City of
Sheboygan Regional Wastewater Treatment
Wausau Water Works

Australia
South Australian Water Corporation
Sydney Water Corporation
Water Corporation of Western Australia

Canada
Greater Vancouver Regional District
Regina, City of, Saskatchewan
Toronto, City of, Ontario
Winnipeg, City of, Manitoba

New Zealand
Watercare Services Limited

United Kingdom
Yorkshire Water Services Limited

STORMWATER UTILITY

California
Los Angeles, City of, Department of Public Works
Monterey, City of
San Francisco, City & County of
Santa Rosa, City of

Colorado
Aurora, City of
Boulder, City of

Georgia
Griffin, City of

Iowa
Cedar Rapids Wastewater Facility
Des Moines, City of

Kansas
Overland Park, City of

Maine
Portland Water District

Minnesota
Western Lake Superior Sanitary District

North Carolina
Charlotte, City of, Stormwater Services

Pennsylvania
Philadelphia, City of

Tennessee
Chattanooga Stormwater Management

Washington
Bellevue Utilities Department
Seattle Public Utilities

STATE

Arkansas Department of Environmental Quality
Fresno Metropolitan Flood Control District
Kansas Department of Health & Environment
Kentucky Department of Environmental Protection
Ohio River Valley Sanitation Commission
Urban Drainage & Flood Control District, CO

CORPORATE

Abt. Associates
ADS Environmental Services
Alan Plummer & Associates
Alden Research Laboratory Inc.
American Water
Aqua-Aerobic Systems Inc.
Aquateam–Norwegian Water Technology Centre A/S
ARCADIS
Associated Engineering
Black & Veatch
Blasland, Bouck & Lee Inc.
Boyle Engineering Corporation
Brown & Caldwell
Burns & McDonnell
CABE Associates Inc.
The Cadmus Group
Camp Dresser & McKee Inc.
Carollo Engineers Inc.
Carpenter Environmental Associates Inc.
CDS Technologies Inc.
Chemtrac Systems Inc.
CH2M HILL
Construction & Environmental Consulting, Inc.
D&B/Guarino Engineers, LLC
Damon S. Williams Associates, LLC
Dewling Associates Inc.
Earth Tech Inc.
Ecovation
EMA Inc.
Environ/The ADVENT Group, Inc.
The Eshelman Company Inc.
EWT Holdings Corporation
Fay, Spofford, & Thorndike Inc.
Freese & Nichols Inc.
ftn Associates Inc.
Fuss & O'Neill Inc.
Gannett Fleming Inc.
Geosyntec Consultants
GHD
Golder Associates Ltd.
Greeley and Hansen LLC
Hazen & Sawyer, P.C.
HDR Engineering Inc.
HNTB Corporation
HydroQual Inc.
Infilco Degremont Inc.
Jacobson Helgoth Consultants Inc.
Jason Consultants LLC Inc.
Jordan, Jones, & Goulding Inc.
KCI Technologies Inc.
Kelly & Weaver, P.C.
Kennedy/Jenks Consultants
KMK Consultants
Komline Sanderson Engineering Corporation
Lawler, Matusky & Skelly Engineers, LLP
Limno-Tech Inc.
Lombardo Associates Inc.
Malcolm Pirnie Inc.
Material Matters
McKim & Creed
MEC Analytical Systems Inc.
Metcalf & Eddy Inc.
MPR Engineering Corporation, Inc.
MWH
New England Organics
O'Brien & Gere Engineers Inc.
Odor & Corrosion Technology Consultants Inc.
Oswald Green, LLC
PA Government Services Inc.
Parametrix Inc.
Parsons
Post, Buckley, Schuh & Jernigan
R&D Engineering/Conestoga Rover & Associates
RMC Inc.
R.M. Towill Corporation
Ross & Associates Ltd.
Rothberg, Tamburini & Windsor, Inc.
Royce Technologies
Stantec Consulting Inc.
Stearns & Wheler, LLC
Stone Environmental Inc.
Stormwater360
Stratus Consulting Inc.
Synagro Technologies Inc.
Tetra Tech Inc.
Trojan Technologies Inc.
Trussell Technologies, Inc.
URS Corporation
USfilter
Wade-Trim Inc.
Westin Engineering Inc.
Weston Solutions Inc.
Woodard & Curran
WRc/D&B, LLC
WWETCO, LLC
Zenon Environmental Inc.
Zoeller Pump Company

INDUSTRY

American Electric Power
ChevronTexaco Energy Research & Technology Company
The Coca-Cola Company
Dow Chemical Company
DuPont Company
Eastman Chemical Company
Eastman Kodak Company
Eli Lilly & Company
Merck & Company Inc.
Premier Chemicals LLC
Procter & Gamble Company
RWE Thames Water Plc
Severn Trent Services Inc.
Suez Environnment
United Water Services LLC

Note: List as of 3/1/06

WERF Board of Directors

Chair
Vernon D. Lucy
Infilco Degremont Inc.

Vice-Chair
Dennis M. Diemer, P.E.
East Bay Municipal Utility District

Secretary
William J. Bertera
Water Environment Federation

Treasurer
James M. Tarpy, J.D.
Metro Water Services

Mary E. Buzby, Ph.D.
Merck & Company Inc.

Mohamed F. Dahab, Ph.D.
University of Nebraska, Lincoln

Glen T. Daigger, Ph.D.
CH2M HILL

Robert W. Hite, J.D.
Metro Wastewater Reclamation District

Jerry N. Johnson
District of Columbia Water & Sewer Authority

Richard D. Kuchenrither, Ph.D.
Black & Veatch Corporation

Alfonso R. Lopez, P.E.
New York City Department of Environmental Protection

Richard G. Luthy, Ph.D.
Stanford University

Lynn H. Orphan, P.E.
Kennedy/Jenks Consultants

Murli Tolaney, P.E., DEE
MWH

Alan H. Vicory, Jr., P.E., DEE
Ohio River Valley Water Sanitation Commission

Executive Director
Glenn Reinhardt

WERF Research Council

Chair
Glen T. Daigger, Ph.D.
CH2M HILL

Vice-Chair
Peter J. Ruffier
Eugene/Springfield Water Pollution Control

Christine F. Andersen, P.E.
City of Long Beach, California

Robert G. Arnold, Ph.D.
University of Arizona

Gail B. Boyd
URS Corporation

William C. Boyle, Ph.D.
University of Wisconsin

William L. Cairns, Ph.D.
Trojan Technologies Inc.

Robbin W. Finch
Boise City Public Works

Ephraim S. King
U.S. EPA

Mary A. Lappin, P.E.
Kansas City Water Services Department

Keith J. Linn
Northeast Ohio Regional Sewer District

Brian G. Marengo, P.E.
City of Philadelphia Water Department

Drew C. McAvoy, Ph.D.
The Procter & Gamble Company

Margaret H. Nellor, P.E.
Nellor Environmental Associates, Inc.

Karen L. Pallansch
Alexandria Sanitation Authority

Steven M. Rogowski, P.E.
Metro Wastewater Reclamation District of Denver

Michael W. Sweeney, Ph.D.
EMA Inc.

George Tchobanoglous, Ph.D.
Tchobanoglous Consulting

Gary Toranzos, Ph.D.
University of Puerto Rico

Ben Urbonas, P.E.
Urban Drainage and Flood Control District

James Wheeler, P.E.
U.S. EPA

WERF Product Order Form

As a benefit of joining the Water Environment Research Foundation, subscribers are entitled to receive one complimentary copy of all final reports and other products. Additional copies are available at cost (usually $10). To order your complimentary copy of a report, please write "free" in the unit price column. WERF keeps track of all orders. If the charge differs from what is shown here, we will call to confirm the total before processing.

Name Title

Organization

Address

City State Zip Code Country

Phone Fax Email

Stock #	Product	Quantity	Unit Price	Total
			Postage & Handling	
			VA Residents Add 4.5% Sales Tax	
			Canadian Residents Add 7% GST	
			TOTAL	

Method of Payment: (All orders must be prepaid.)

❑ Check or Money Order Enclosed

❑ Visa ❑ Mastercard ❑ American Express

Account No. Exp. Date

Signature

Shipping & Handling:

Amount of Order	United States	Canada & Mexico	All Others
Up to but not more than:	Add:	Add:	Add:
$20.00	$5.00*	$8.00	50% of amount
30.00	5.50	8.00	40% of amount
40.00	6.00	8.00	
50.00	6.50	14.00	
60.00	7.00	14.00	
80.00	8.00	14.00	
100.00	10.00	21.00	
150.00	12.50	28.00	
200.00	15.00	35.00	
More than $200.00	Add 20% of order	Add 20% of order	

*minimum amount for all orders

Note: Please make checks payable to the Water Environment Research Foundation.

To Order (Subscribers Only):

Log on to www.werf.org and click on the "Product Catalog."

Phone: (703) 684-2470
Fax: (703) 299-0742.

WERF
Attn: Subscriber Services
635 Slaters Lane
Alexandria, VA 22314-1177

To Order (Non-Subscribers):

Non-subscribers may be able to order WERF publications either through WEF (www.wef.org) or IWAP (www.iwapublishing.com). Visit WERF's website at www.werf.org for details.

IWA Publishing's authorised EU representative for General Product Safety Regulations is Diane D'Arras, 15 rue Duret, 75116 Paris, France, e-mail: safety@iwap.co.uk.

Printed and bound by CPI Group (UK) Ltd, Croydon, CR0 4YY
12/05/2026
02108876-0015